RAND | NATIONAL DEFENSE RESEARCH INSTITUTE

CURRENT AND FUTURE CHALLENGES TO

Resourcing U.S. Navy Public Shipyards

Jessie Riposo
Michael E. McMahon
James G. Kallimani
Daniel Tremblay

For more information on this publication, visit www.rand.org/t/RR1552

Library of Congress Cataloging-in-Publication Data is available for this publication.
ISBN: 978-0-8330-9762-0

Published by the RAND Corporation, Santa Monica, Calif.

© Copyright 2017 RAND Corporation

RAND® is a registered trademark.

Cover: composite of U.S. Navy photos; mechanic photo by Wendy Hallmark.

Support RAND
Make a tax-deductible charitable contribution at
www.rand.org/giving/contribute

www.rand.org

Preface

Managing the U.S. Navy's four public shipyards is a challenge. The shipyards must be ready and able to support the fleet anytime and anywhere in the world at a moment's notice. For this reason, the public shipyards are required to maintain core capabilities in ship repair that the private sector does not maintain. In addition, they are subject to laws and regulations that dictate how and where work can be performed.

Between fiscal years 2004 and 2014, the number of civilians employed at the Navy's four public shipyards increased by 17 percent, while the direct man-days executed increased by just 7 percent. The significant increase in personnel without a comparable increase in workload has raised many questions, and possible explanations include changes in the amount and type of work, workforce composition, performance during execution, and organizational goals, such as desired responsiveness.

To better understand the reasons for these trends, the Office of the Chief of Naval Operations, Assessment Division, asked the RAND National Defense Research Institute to identify influences on Navy shipyard manning requirements, examine the near- and middle-term challenges in planning and programming for these workforce resources, and provide recommendations for improving resource planning for the shipyards. In turn, RAND researchers explored trends in Navy shipyard work, whether more personnel are now required to perform it, and what risks shipyards may wish to address through additional hiring. This study was exempt from human subjects protection because it did not include analysis of discernable individual human subjects, only the workforce in aggregate.

This research should be of interest to the Navy's senior leadership and operations leadership, public shipyard planners, Naval Sea System Command, and others more generally interested in public shipyards, resource planning, and workforce planning.

This research was conducted within the Acquisition and Technology Policy Center of the RAND National Defense Research Institute, a federally funded research and development center sponsored by the Office of the Secretary of Defense, the Joint Staff, the Unified Combatant Commands, the Navy, the Marine Corps, the defense agencies, and the defense Intelligence Community.

For more information on the RAND Acquisition and Technology Policy Center, see www.rand.org/nsrd/about/atp or contact the director (contact information is provided on the web page).

Contents

Figures

Tables

Summary

The U.S. Navy currently owns and operates four public shipyards, which must be ready and able to support the fleet anytime and anywhere in the world at a moment's notice. They perform the Navy's most-complex maintenance and modernization, including for nuclear-powered submarines and aircraft carriers. For this reason, the public shipyards are required to maintain core capabilities that the private sector does not maintain. In addition, they are subject to laws and regulations that dictate how and where work can be performed.

Over the past five years, workload at the Navy's public shipyards has been on the rise. Direct man-days of work assigned to and executed by the shipyards have increased during that time and are planned to continue to increase in the near future.[1] Indirect man-days have also risen. Some of these increases have been driven by the introduction of new classes of platforms maintained at the shipyards, more work for aging classes of carriers and submarines, and higher operational tempo. Increases in programmed work for nuclear-powered ballistic-missile submarines (SSBNs) and execution challenges in that work have driven additional workload. Moreover, loss of productivity from the "greening" of the workforce—that is, an influx of new, and thus inexperienced, personnel—has slowed productivity and will continue to do so in the near and middle terms. Navy initiatives to more

---

[1] *Direct man-days* are man-days associated directly with a project. They include wrench turning, welding, and other production activities. *Indirect man-days* include training, supervision, and administrative activities, such as finance or human resources activities, that benefit all projects.

rapidly train the newly hired trade personnel have shown early success and may play a key role in future workforce management as the initiatives are broadened. Planned increases in civilian staffing levels are necessary but not sufficient to mitigate near-term execution risk at the shipyards.

Based on historical data and forecasts in the shipyards' Workload Allocation and Resource Reports (WARRs) provided to the project team by each shipyard, civilian staffing levels have outpaced workload increases at the four public shipyards in recent years and are expected to continue to do so. More specifically, from fiscal years (FYs) 2004 through 2014, the number of full-time equivalent civilian employees increased by 17 percent while direct man-days increased by just 7 percent, as shown in Figure S.1. Between FYs 2004 and 2018, civilian staffing will increase by 34 percent and direct man-days will increase by 29 percent, representing a closer alignment between people and work. Total man-days (direct and indirect) will increase by 33 percent between FYs 2008 and 2018.[2]

To better understand the causes of the increases in civilian personnel, the Office of the Chief of Naval Operations, Assessment Division, asked the RAND National Defense Research Institute to help identify influences on the manning requirements for naval shipyards, examine the near- and middle-term challenges in planning and programming for these workforce resources, and provide recommendations for improving resource planning for the shipyards.

Discussions with stakeholders and research on workforce management pointed to the following three research questions to guide our

---

[2] Norfolk Naval Shipyard, "WF-300 Workload Allocation and Resource Report (WARR)," spreadsheet, provided to RAND by the Naval Sea Systems Command, Logistics, Maintenance, and Industrial Operations Directorate (NAVSEA 04), July 2014; Pearl Harbor Naval Shipyard, "WF-300 Workload Allocation and Resource Report (WARR)," spreadsheet, provided to RAND by NAVSEA 04, July 2014; Portsmouth Naval Shipyard, "WF-300 Workload Allocation and Resource Report (WARR)," spreadsheet, provided to RAND by NAVSEA 04, July 2014; and Puget Sound Naval Shipyard, "WF-300 Workload Allocation and Resource Report (WARR)," spreadsheet, provided to RAND by NAVSEA 04, July 2014.

Figure S.1
Civilian Staffing Levels and Man-Days Executed and Planned at Public Shipyards, FYs 2004–2018

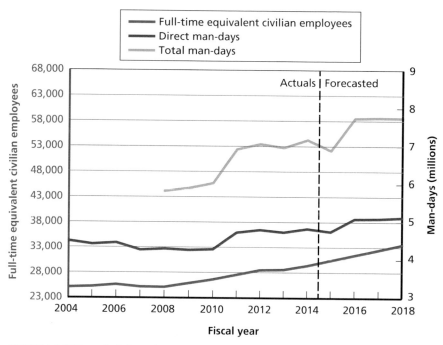

SOURCE: RAND analysis based on Norfolk Naval Shipyard, 2014; Pearl Harbor Naval Shipyard, 2014; Portsmouth Naval Shipyard, 2014; and Puget Sound Naval Shipyard, 2014.
NOTE: Indirect man-days prior to FY 2008 are not provided because the data were not available. Prior to 2007, the shipyards were transitioning from the Navy's Working Capital Fund to the mission funding model. Starting in FY 2007, all shipyards became mission-funded.
RAND RR1552-S.1

assessment of why civilian staffing levels have increased more rapidly than workload:

- How is shipyard work changing?
- Are more personnel now required to perform the same work?
- What are the shipyards' root issues and risks that require additional hiring?

We review our findings for each question next.

Increases in Workload

Workload has increased at the shipyards, particularly for SSBNs and nuclear-powered aircraft carriers (CVNs), as well as for other scheduled maintenance. In FY 2000, the naval shipyards supported a fleet of eight CVNs;[3] by FY 2016, they supported 11. Over a CVN's life cycle, the shipyards will perform about 7.5 million total man-days of maintenance (excluding the refueling and complex overhaul). This alone results in an average annual increase of a little more than 150,000 man-days per ship, or 450,000 annual man-days across the three additional carriers the shipyards were supporting beginning in FY 2016.

Operating and maintenance cycles have also changed in ways that could increase workload. The operational cycle—including deployment, training, and maintenance—increased from 24 months in FY 2004 to 32 months in FY 2006, and it is expected to increase to 36 months in the Optimized Fleet Response Plan (OFRP).[4] One result of these increases has been a need to accomplish more work in fewer maintenance periods, referred to as *availabilities*. In 2004, there were 25 availabilities for nuclear carriers; in FY 2013, there were 18; and under the OFRP, there are expected to be only 16. This leads to larger, less-frequent availabilities, which are more difficult for the shipyards to manage. The consequence of increased peaks and lower valleys in workload is an increased likelihood of inefficiency in execution.

In FY 2004, the Class Maintenance Plan for the SSBN fleet required approximately 406,000 man-days of maintenance over the life of each boat (or 6.5 million man-days for the fleet of 16 boats). In FY 2007, this increased to 459,000 man-days of maintenance over the life of each boat (or 7.3 million man-days for the fleet). For the fleet of 16 boats, this equates to a base increase in work of approximately 280,000 man-days per year between FYs 2004 and 2007. In addition to the workload increasing, the largest maintenance period

---

[3]   Based on shipyard WARR files (Norfolk Naval Shipyard, 2014; Pearl Harbor Naval Shipyard, 2014; Portsmouth Naval Shipyard, 2014; and Puget Sound Naval Shipyard, 2014).

[4]   The carriers will not all be on the OFRP at the same time, because their maintenance plans roll over from the old plan to the OFRP at different points.

for the SSBN, the engineered refueling overhaul (ERO), has increased in duration from approximately 28 months to 33 months. Those who are involved with executing this work believe the increase in duration is a result of executing with insufficient resources. The work at the shipyards is prioritized in a way that results in the longer ERO process being delayed as limited resources are diverted to complete other availabilities. One result of this increase in ERO duration is an increase in the indirect costs associated with the availability. The public shipyards use a *direct labor index*—that is, the proportion of total man-days attributed to direct labor—to plan indirect man-days as a function of direct man-days. From FYs 2004 through 2014, the direct labor index increased from 51 percent to 57 percent; therefore, an increase by one direct man-day will be accompanied by an increase of approximately 0.4 indirect man-days.[5]

Scheduled maintenance at the shipyards has also increased significantly. Work falling into this category includes oversight of private-sector activities under the purview of the shipyard, continuous maintenance activity, ship alterations, nuclear equipment disposal, fleet maintenance availabilities, Nuclear Regional Maintenance Department activities, fleet technical support, availability planning activities, and process activities. Increases may have occurred as a result of the lengthening operational cycle, which may have pushed more work into continuous maintenance, fleet maintenance availabilities, and fleet technical support activities.

Indirect man-days have also increased as a result of the increases in direct work, albeit at a greater rate than direct man-days have increased. Contributors to increases in indirect man-days include training of new workers and additional regulatory and policy burdens. Better data are needed to quantify the effects of these additional causes.

Based on forecasted data in the WARRs, the future workload plan shows a near-term increase in expected work, with workload peaking in 2018 at a level that will be 33 percent higher than in 2010. This significant increase in work includes maintenance activity that has never been performed, such as that required for the Navy's newest attack submarine.

---

[5]   These figures were provided to the study team by NAVSEA 04.

Changing Workforce Demographics

As workload has increased, workforce experience has decreased. For example, the percentage of the total civilian workforce with less than ten years of experience has increased from 35 percent in FY 2006 to nearly 50 percent in FY 2014, while the percentage with 20–29 years of experience decreased from 31 percent to 12 percent. Figure S.2 shows the changing composition of the workforce in recent years.[6]

As the proportion of the workforce with little to no experience rises, total output is expected to decline. RAND researchers developed a simple model to explore the implications of changing demographics to workforce productivity.[7] The model estimates the future *predicted need*, which is the number of civilians that would have to be hired to achieve the maximum productive workforce, and the *effective workforce*, which is the number of fully productive workers. In each year, the model hires as many people as possible, within the constraints of practical hiring, until the effective workforce reaches the "planned" workforce level. The *planned workforce* is what is currently in the Navy's budget and Program Objective Memorandum (POM) for the shipyards. At the time of this analysis, the goal, set by the public shipyards, is to reach 33,500 civilian staff by FY 2017.

As the total shipyard workforce increases to 33,500 and experienced workers are replaced with less-experienced ones, we observe a decline in overall workforce productivity. This assumes that a first-year apprentice is one-fourth as productive as a fully experienced journeyman, meaning that four first-year apprentices would be needed to replace the work of one experienced journeyman.[8] While some lost productivity can be recovered through additional hires, not all can. The productivity deficit is such that the shipyards cannot hire and train

---

[6]   Norfolk Naval Shipyard, 2014; Pearl Harbor Naval Shipyard, 2014; Portsmouth Naval Shipyard, 2014; and Puget Sound Naval Shipyard, 2014.

[7]   See Jessie Riposo, Brien Alkire, John F. Schank, Mark V. Arena, James G. Kallimani, Irv Blickstein, Kimberly Curry Hall, and Clifford A. Grammich, *U.S. Navy Shipyards: An Evaluation of Workload- and Workforce-Management Practices*, Santa Monica, Calif.: RAND Corporation, MG-751-NAVY, 2008.

[8]   Riposo et al., 2008.

Figure S.2
Civilian Workforce Experience, FYs 2006–2014

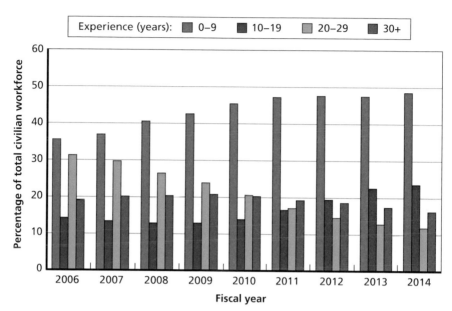

SOURCE: Norfolk Naval Shipyard, 2014; Pearl Harbor Naval Shipyard, 2014;
Portsmouth Naval Shipyard, 2014; and Puget Sound Naval Shipyard, 2014.
RAND RR1552-S.2

the numbers of people that would be required to replace lost productivity to meet near-term peak demands. Nevertheless, if the yards hire in addition to the current plan, they can achieve an effective workforce that meets the planned workforce requirement by FY 2017.

Mitigation for Future Challenges and Risks

Forecasting the future work that the naval shipyards will have to perform, which is a primary component in workforce planning, is a challenging task. Budgets for the number of man-days required to perform depot-level maintenance at the public shipyards are programmed two years ahead of the execution year, and, for a variety of reasons, those budgets are consistently below what is ultimately required in the execution year. This is a result of factors that the budgeting has not accounted for, such as operational cycles, unidentified maintenance, time between maintenance, unplanned events, and age of the fleet. Accounting for this historical trend, as well as current workload forecasts, may mean that future peak workload could be nearly 50 percent greater than that in FY 2010.

As the Navy adopts new operational cycles that have increased deployment durations and that increase the amount of time between major availabilities, unplanned maintenance events are likely to occur. In addition, the increasing average age of all nuclear vessels maintained at the Navy shipyards suggests that there likely will be additional unplanned maintenance. There are also new lines of work that the shipyards have less experience performing and for which the maintenance plans are still evolving. This includes the maintenance required to support *Virginia*-class submarines, *Ford*-class CVNs, and CVN inactivation. These risks will materialize during a time of workforce transition toward less experience at the public shipyards. The workforce will begin to gain productivity over time, but the confluence of a peak workload, new lines of work, and an inexperienced workforce represents risk that must be carefully managed and resourced.

Conclusions

Although naval shipyard manning levels have been increasing more than recent workloads and recent end-strength additions have been approved and programmed for the naval shipyards, these increases will not suffice to fulfill peak demands through FY 2018. The shipyards also face a productivity deficit created by the increasingly inexperienced workforce—a deficit that cannot be overcome solely through hiring. The Navy is examining near-term steps to outsource some maintenance availabilities to the private sector for short-duration submarine availabilities. Navy leaders will need to undertake additional strategies, such as more outsourcing of work, including possibly privatizing work for some of the planned inactivation and recycling workloads for carriers and submarines. In order to build an effective workforce able to meet the planned requirements, the shipyards will need to hire more civilians. Until these individuals gain productivity, the effective workforce will not meet the planned requirement. But when they do gain productivity, the effective workforce will meet, and then exceed, planned workforce levels. Such excess capacity will not likely appear before 2023, but it might then suffice to help the shipyards manage unpredictable fluctuations in future workload, although further investigation is required to determine this effect.

Acknowledgments

This report could not have been completed without the guidance and assistance of many individuals. We wish to thank the team at the Office of the Chief of Naval Operations (OPNAV), Assessment Division (N81), which sponsored this project, for providing feedback throughout. Those individuals include Arthur Barber, Charles Werchado, RADM Herman Shelanski, CAPT Greg Sheahan, CDR Neil Sexton, Carlton Hill, and Steven Williams. We also wish to thank RDML Robert Burke, CAPT John Lobouno, and Guy Holsten from OPNAV, Fleet Readiness Division (N43), for their support and insight. Naval Sea Systems Command (NAVSEA), Logistics, Maintenance, and Industrial Operations Directorate (04), provided the research team with valuable data and insight into the naval shipyards. We thank RADM Mark Whitney, Jim Wreski, Hank Zajic, and Larry Marquess for their continued support of this project. We also thank VADM William Hilarides and William Deligne at NAVSEA Headquarters for providing their support during the research process. The program managers for the In-Service Aircraft Carrier Program and the Strategic and Attack Submarine Program (PMS 312 and PMS 392, respectively), as well as the Carrier Planning Activity and the Submarine Maintenance Engineering Planning and Procurement Activity, provided very useful data and insightful feedback during the project.

We also wish to acknowledge the time provided by the Fleet. At U.S. Fleet Forces Command N43, we thank RADM Richard Berkey and Stephanie Douglas. From Commander, Submarine Forces Atlantic (N43), we thank CAPT Michael Temme. From Commander, Naval

Air Forces Atlantic (N43), we thank CAPT Douglas Lemon. From Commander, Pacific Fleet (N43), we thank RDML Alma Grocki, Ken Voorhees, and CAPT Alex Desroches.

We wish to thank the shipyard leadership and personnel. With the participation of Norfolk Naval Shipyard, Puget Sound Naval Shipyard and Intermediate Maintenance Facility, Portsmouth Naval Shipyard, and Pearl Harbor Naval Shipyard and Intermediate Maintenance Facility, we were able to perform higher-quality research. Each shipyard arranged research visits, built data, and spent valuable time corresponding with the research team.

We thank Irv Blickstein of the RAND Corporation and CAPT Mike Ford from the U.S. Navy for their thorough and helpful reviews of our work.

From RAND, this report benefited from a number of contributions. In particular, Jamie Greenberg provided administrative support, and Debbie Peetz provided document and literary search support.

Abbreviations

CNO	Chief of Naval Operations
CVN	nuclear-powered aircraft carrier
DPIA	docking planned incremental availability
ERO	engineered refueling overhaul
ERP	extended refit period
FY	fiscal year
IMP	Incremental Maintenance Plan
IRR	inactivation, reactor compartment disposal, recycling
MTS	moored training ship
N43	Fleet Readiness Division
NAVSEA	Naval Sea Systems Command
NAVSEA 04	Naval Sea Systems Command, Logistics, Maintenance, and Industrial Operations Directorate
NNSY	Norfolk Naval Shipyard
OFRP	Optimized Fleet Response Plan
OPNAV	Office of the Chief of Naval Operations

PHNSY	Pearl Harbor Naval Shipyard
PIA	planned incremental availability
PNSY	Portsmouth Naval Shipyard
POM	Program Objective Memorandum
PSA	post-shakedown availability
PSNSY	Puget Sound Naval Shipyard
RCOH	refueling and complex overhaul
SSBN	nuclear-powered ballistic-missile submarine
SSGN	nuclear-powered cruise-missile submarine
SSN	nuclear-powered attack submarine
WARR	Workload Allocation and Resource Report

CHAPTER ONE

Introduction

The U.S. Navy currently owns and operates four public shipyards: Norfolk Naval Shipyard (NNSY) in Norfolk, Virginia; Pearl Harbor Naval Shipyard (PHNSY) and Intermediate Maintenance Facility in Pearl Harbor, Hawaii; Portsmouth Naval Shipyard (PNSY) in Kittery, Maine; and Puget Sound Naval Shipyard (PSNSY) and Intermediate Maintenance Facility in Bremerton, Washington. The Fleet Commanders determine how the shipyards' resources are employed, and the Naval Sea Systems Command (NAVSEA) operates and manages the shipyards.

These shipyards perform the most-complex maintenance that the Navy requires, including most depot-level and some intermediate-level life-cycle maintenance and modernization of nuclear-powered ballistic-missile submarines (SSBNs), nuclear-powered cruise-missile submarines (SSGNs), nuclear-powered attack submarines (SSNs), and nuclear-powered aircraft carriers (CVNs). They also perform refueling of SSNs and SSBNs; life-cycle sustainment and refueling of moored training ships (MTSs, which currently are all former SSBNs, although the next MTSs to be converted will be retired SSNs); and inactivation, reactor compartment disposal, recycling (IRR) of SSNs, SSBNs, and CVNs. The shipyards are also home to regional repair centers, which provide planning yard functions, intermediate-level maintenance on equipment, maintenance of key national-security infrastructure, and systems maintenance and modernization for special projects.[1]

The naval shipyard mission has evolved and expanded in the past decade. The shipyards are now responsible for managing and executing with broad regional maintenance responsibilities.[2] This means that the shipyards are now responsible for not only the work occurring within their gates but also any maintenance work occurring within the same region at other privately owned shipyards. They provide management and oversight of work that is contracted out to the private sector.[3] This burden generates increased manpower demand.

A decade ago, Navy shipyards maintained eight CVNs; by FY 2016, they maintained 11. Carrier and submarine operating cycles have also changed, resulting in longer periods of time between sched-

---

[1] Key infrastructure vital to national security is embedded and maintained at the four Navy public shipyards. These facilities include the only government-owned dry docks capable of docking a nuclear aircraft carrier and certified for docking nuclear carriers and submarines. Additionally, the naval shipyards have deep-water berths, piers, and wharfs for U.S. Navy ships and submarines and large gantry and portal cranes certified for nuclear maintenance. The four shipyards also contain unique machine-shop plant equipment and facilities required for maintenance of the Navy's capital vessels.

[2] Two of the shipyards—Puget Sound Naval Shipyard and Intermediate Maintenance Facility and Pearl Harbor Naval Shipyard and Intermediate Maintenance Facility—have integrated the regional maintenance activities. At Puget Sound, the Intermediate Maintenance Facility at Naval Submarine Base Bangor and Intermediate Maintenance Activity at Naval Station Everett were integrated fully into the naval shipyard in 2002. The Bangor facility performed maintenance and modernization on *Trident*-class SSBNs, and the Everett activity performed intermediate-level maintenance on home-ported surface ships at Naval Station Everett. The Puget Sound and Pearl Harbor shipyards now include a larger, fully integrated regional fleet maintenance and modernization execution and oversight role, in addition to oversight and contracting of private-sector work within shipyard-led availabilities.

[3] The Navy's public shipyards are designated as Lead Maintenance Activities for the fleet maintenance availabilities that they plan and perform. As such, the shipyards are responsible to the Fleet, via the Type Commander and NAVSEA, for final certification of work completion for all maintenance performed. This includes private-sector work performed in these availabilities, which requires the Navy shipyards to integrate all work into an executable and safe overall plan and to maintain oversight of work process controls. Additionally, the two shipyards that have integrated regional Intermediate Maintenance Facilities into the shipyard (Puget Sound and Pearl Harbor) have a contracting role in overseeing work performed by the private sector under a multi-ship, multi-option or other contracting vehicle.

uled maintenance. In addition, Navy shipyards have adopted several significant lines of new work in recent decades, including maintenance of *Virginia*-class SSNs, MTS conversions, and CVN inactivation planning. As the Navy shifts more resources to the Pacific, the shipyards have become increasingly responsible for forward-deployed naval force maintenance. They are currently supporting a nuclear aircraft carrier in Yokosuka, Japan, and several forward-based SSNs in Guam, and they respond to emergent repairs all over the world. Furthermore, the way the public shipyards are funded has changed, which affects how they operate. They transitioned from the Navy's Working Capital Fund model, in which the yard was reimbursed for each service performed, to mission funding, whereby a certain capacity is purchased up front and then allocated as the year progresses.

The Problem

These increasing demands have led to a greater number of direct man-days of work at the shipyards, as Figure 1.1 illustrates.[4] However, the number of full-time equivalent civilian workers in the shipyards has increased at a higher pace recently. These numbers are based on workload plans, referred to as Workload Allocation and Resource Reports (WARRs), provided to the project team by each shipyard.

Between fiscal years (FYs) 2004 and 2014, the number of civilian workers at the public shipyards increased by 17 percent while the direct man-days executed increased by just 7 percent.[5] From FYs 2007 through 2014, the number of military personnel increased by approximately 500, while civilian staff increased by approximately 6,200. Civilian employment at the public shipyards can be seen in Figure 1.2.

---

[4] *Direct man-days* are man-days associated directly with a project. They include wrench turning, welding, and other production activities. *Indirect man-days* include training, supervision, and administrative activities, such as finance or human resources activities, that benefit all projects.

[5] The direct man-days do not include work that is planned to be contracted out.

Figure 1.1
Civilian Staffing Levels and Man-Days Executed and Planned at Public Shipyards, FYs 2004–2018

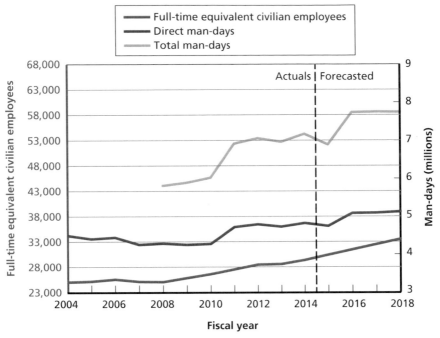

SOURCE: Norfolk Naval Shipyard, "WF-300 Workload Allocation and Resource Report (WARR)," spreadsheet, provided to RAND by NAVSEA 04, July 2014; Pearl Harbor Naval Shipyard, "WF-300 Workload Allocation and Resource Report (WARR)," spreadsheet, provided to RAND by NAVSEA 04, July 2014; Portsmouth Naval Shipyard, "WF-300 Workload Allocation and Resource Report (WARR)," spreadsheet, provided to RAND by NAVSEA 04, July 2014; and Puget Sound Naval Shipyard, "WF-300 Workload Allocation and Resource Report (WARR)," spreadsheet, provided to RAND by NAVSEA 04, July 2014.
NOTE: Indirect man-days prior to FY 2008 are not provided because the data were not available. Prior to 2007, the shipyards were transitioning from the Navy's Working Capital Fund to the mission funding model. Starting in FY 2007, all shipyards became mission-funded.
RAND RR1552-1.1

In addition, the employment of U.S. military personnel at the public shipyards has increased slightly, as shown in Figure 1.3. The proportion of military to civilian staff has remained constant, at around 7 percent of the total force.

Figure 1.2
Civilian Employees at Public Shipyards, FYs 2004–2017

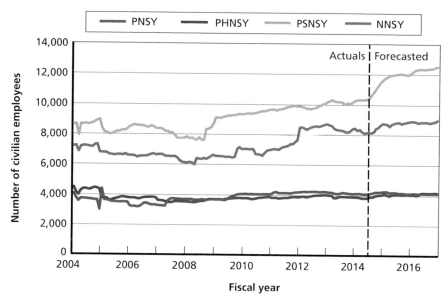

SOURCE: Norfolk Naval Shipyard, 2014; Pearl Harbor Naval Shipyard, 2014;
Portsmouth Naval Shipyard, 2014; and Puget Sound Naval Shipyard, 2014.
RAND RR1552-1.2

Between FYs 2004 and 2014, overtime levels declined, but in most cases, they are still well above a cost effective level (see Figure 1.4).[6] This suggests that a disproportionate increase in civilian personnel would be desirable. However, it is expected that the increases in direct man-days and civilian staffing levels from FYs 2014 through 2018 will begin to level out. Based on historical and forecasted data provided by the shipyards, civilian staffing will increase by 34 percent during that

---

[6]   See Jessie Riposo, Brien Alkire, John F. Schank, Mark V. Arena, James G. Kallimani, Irv Blickstein, Kimberly Curry Hall, and Clifford A. Grammich, *U.S. Navy Shipyards: An Evaluation of Workload- and Workforce- Management Practices*, Santa Monica, Calif.: RAND Corporation, MG-751-NAVY, 2008. This report identifies that use of overtime levels in excess of 12 percent is more costly than using additional permanent employees to complete work. Overtime data were provided to RAND by the NAVSEA Logistics, Maintenance, and Industrial Operations Directorate (NAVSEA 04).

Figure 1.3
Military Employees at Public Shipyards, FYs 2007–2015

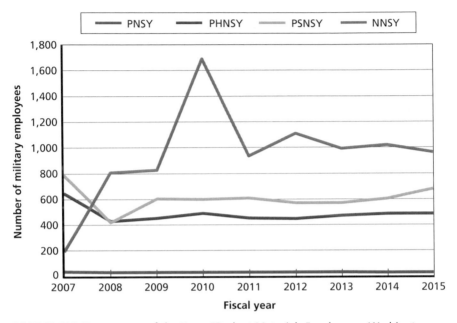

SOURCE: U.S. Department of the Navy, "Budget Materials," web page, Washington, D.C., FYs 2007–2015.
NOTE: Data from President's budget exhibits are available only from FY 2007 forward. Prior to 2007, the shipyards were transitioning from the Navy's Working Capital Fund to the mission funding model. Starting in FY 2007, all shipyards became mission-funded.
RAND *RR1552-1.3*

span, and direct man-days will increase by 29 percent, representing a closer alignment between people and work.

The recent difference between the substantial increase in civilian personnel and the smaller increase in workload has raised many questions, with many potential answers. We explore four possible explanations for this difference, including changes in the amount and type of work, workforce composition, performance during execution, and organizational goals, such as desired responsiveness.[7] To understand

---

[7] For more information on determining workforce levels, see Thomas Bechet, *Strategic Staffing*, New York: American Management Association International, 2002.

Figure 1.4
Overtime Levels Worked at Shipyards, FYs 2004–2013

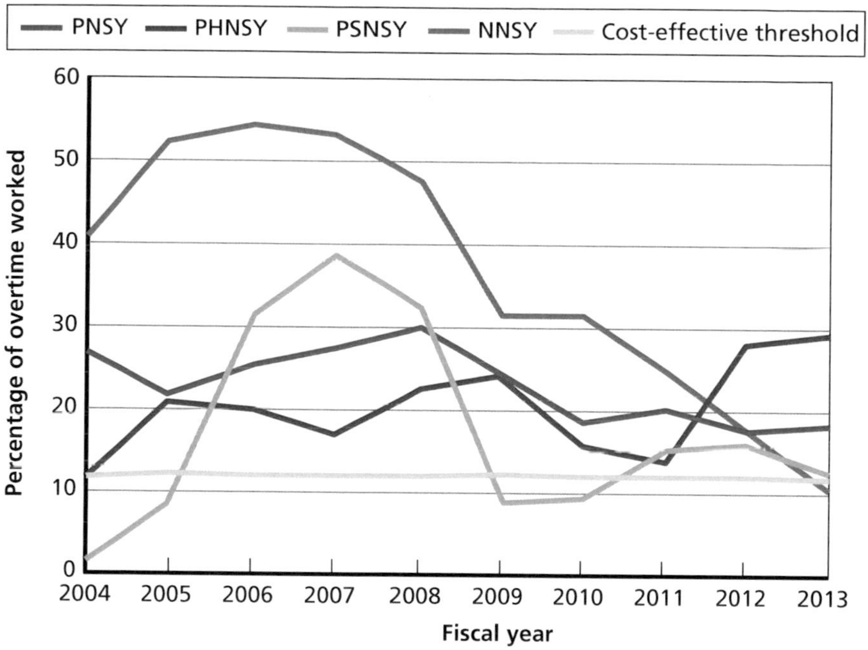

SOURCE: Norfolk Naval Shipyard, 2014; Pearl Harbor Naval Shipyard, 2014;
Portsmouth Naval Shipyard, 2014; and Puget Sound Naval Shipyard, 2014.
RAND RR1552-1.4

how productivity at the shipyards affects resource requirements, we
can develop measures of output and compare them with the resources
required to generate that output. This strategy points to the following
three research questions:

- How is shipyard work changing?
- Are more personnel now required to perform the same work?
- What are the shipyards' root issues and risks that require addi-
  tional hiring?

The study sponsor, the Office of the Chief of Naval Operations
(OPNAV), Assessment Division (N81), asked the RAND National

Defense Research Institute to help identify influences on the manning requirements for naval shipyards, examine the near- and middle-term challenges in planning and programming for these workforce resources, and provide recommendations for improving resource planning for the shipyards. This research should help the Navy identify influences on workforce demand and validate future resource requirements.

Approach

Our analytic approach was organized into five tasks. First, we met with a broad range of key stakeholders—including personnel at NAVSEA 04; OPNAV Fleet Readiness and Logistics Directorate (N4); U.S. Fleet Forces Command (from OPNAV Fleet Readiness Division [N43]); Commander, Pacific Fleet (N43); Commander, Submarine Forces (N43); and others—to characterize the resource-allocation problems and identify possible causes for the discrepancy between the increases in direct man-days and civilian end strength.[8] We then collected a variety of data from NAVSEA 04 and each public shipyard to answer our research questions. These data included, but were not limited to, Department of the Navy budget exhibits, WARRs, workforce demographics, platform-maintenance requirements (as expressed by OPNAV Notice 4700 and Technical Foundation Papers), and maintenance policy. Meetings with leaders and department heads from all four naval shipyards provided opportunities to collect additional data and context. We developed mathematical models and other analytical tools to test our hypotheses and quantify the effect of potential causes of the difference in workforce and workload increases. Finally, we summarized the results of these tasks and provided recommendations.

---

[8] We used our discussions with these stakeholders to inform our overall approach to this analysis. Although we sometimes refer to information that we gleaned from these discussions, we do not cite specific groups or individuals.

Organization of This Report

To understand personnel requirements, we must first understand maintenance requirements. The second chapter of this report provides an overview of the maintenance requirements that the shipyards execute, and we identify where increases in maintenance and maintenance requirements have occurred. Chapter Three describes the analyses for identifying resulting requirements for civilian staffing levels. It highlights the key workforce challenges facing shipyard managers and identifies the implications of a changing workforce demographic. Chapter Four describes some of the future challenges to identifying resource requirements that shipyard managers and the Navy will face. Chapter Five summarizes our findings and provides recommendations for improving resource planning at the public shipyards.

CHAPTER TWO

How Is Shipyard Work Changing?

In FY 2004, the Navy's public shipyards executed nearly 4.5 million direct man-days. The direct work included that for *availabilities*—that is, scheduled assignment of a ship for repairs or modernization, as directed by the Chief of Naval Operations (CNO)—for SSN, SSBN, SSGN, and CVN vessels, as well as some maintenance activities for (non-CVN) surface ships. It also included non-CNO work associated with maintaining these platforms, such as intermediate-level maintenance, continuous-maintenance availabilities and engineering, and design and planning services. Direct man-days began to increase in FY 2010, reached nearly 5 million in FY 2014, and are projected to reach 5.5 million in FY 2018 (see Figure 2.1).

Figure 2.2 shows the number of availabilities under way per month across the four public shipyards from FYs 2008 through 2016 for the ship classes that are supported and for IRR availabilities. The number of SSN availabilities is expected to decline while the number of CVN and IRR availabilities is expected to increase.

To better understand the types of work causing increases in workload, we analyzed man-day trends for each platform and availability type. To discern statistically significant changes in work over time, we analyzed actual man-days executed as published in the shipyards' WARRs. We sought to identify effects on workload within each shipyard, within each class of ship maintained, and by class of ship and shipyard combinations.

Figure 2.1
Direct Man-Days Executed and Planned at Public Shipyards, FYs 2004–2020

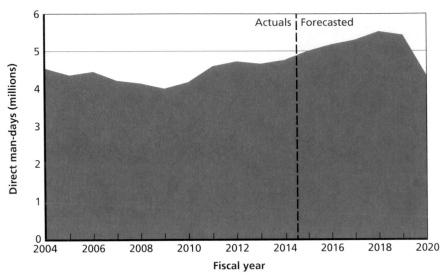

SOURCE: Norfolk Naval Shipyard, 2014; Pearl Harbor Naval Shipyard, 2014;
Portsmouth Naval Shipyard, 2014; and Puget Sound Naval Shipyard, 2014.
RAND RR1552-2.1

For example, Figure 2.3 shows the actual and projected direct man-days executed, by month, for aircraft carrier availabilities at Puget Sound and Norfolk Naval Shipyards from FYs 2004 through 2016. We applied an autoregressive moving average, a popular statistical tool for modeling that regularly measures data across time and assumes that observations measured more frequently are more correlated than those measured less frequently.[1] We applied this statistical tool to confirm or identify observed trends in the historical data. Figure 2.3 also shows the linear representations of each trend to provide a better idea of the overall trends over time, with Puget Sound increasing and Norfolk slightly decreasing

---

[1]   This is a flexible statistical technique that accounts for the correlation between time points. Our models allow for a simple linear trend across time and include a season effect for the decrease in man-days worked around holidays (e.g., in December). This approach captures the correlation between the number of man-days in sequential months. The assumptions required for use of this tool were met.

Figure 2.2
Number of Availabilities Under Way per Month for Supported Ship Classes and Inactivation, FYs 2008–2016

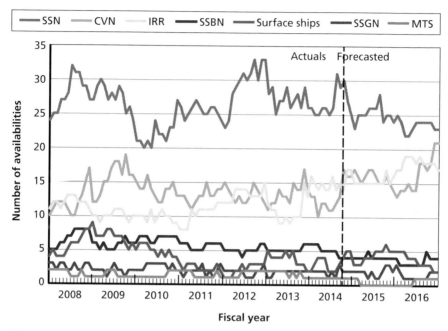

SOURCE: Norfolk Naval Shipyard, 2014; Pearl Harbor Naval Shipyard, 2014; Portsmouth Naval Shipyard, 2014; and Puget Sound Naval Shipyard, 2014.
RAND RR1552-2.2

This analysis indicates that the increase in the overall number of direct man-days across all shipyards is statistically significant ($p < 0.10$), with the greatest increases resulting from SSBN CNO availabilities, CVN CNO availabilities, and scheduled maintenance.[2] Table 2.1 identifies the shipyards and types of work with the most-significant increases in direct man-days between FYs 2004 and 2014.[3]

---

[2] For a description of p-value statistical significance in industrial studies, see Institute for Work and Health, "What Researchers Really Mean by . . . Statistical Significance," *At Work*, No. 40, Spring 2005.

[3] The null hypothesis is that there is no increase in direct man-days. The p-value indicates that the null hypothesis is rejected in favor of the alternative, which is that there is an increase in man-days.

Figure 2.3
Direct Man-Days Executed and Planned per Month for CVN Maintenance at Puget Sound and Norfolk Naval Shipyards, FYs 2004–2016

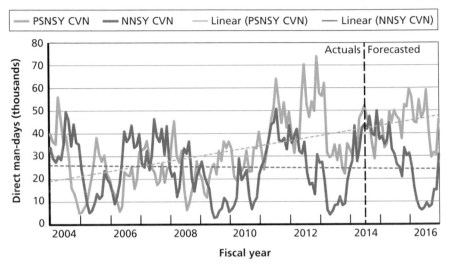

SOURCE: Norfolk Naval Shipyard, 2014; Pearl Harbor Naval Shipyard, 2014; Portsmouth Naval Shipyard, 2014; and Puget Sound Naval Shipyard, 2014.
RAND RR1552-2.3

Table 2.1
Types of Work with the Most-Significant Increases in Direct Man-Days, FYs 2004–2014

Naval Shipyard	Type of Work	P-Value	Increase in Direct Man-Days
Puget Sound, Norfolk	SSBN CNO availabilities	0.00	346,902
Puget Sound, Norfolk, Portsmouth	Scheduled maintenance	0.00	202,196
Norfolk	MTS and MTS conversions	0.05	49,704
Puget Sound	CVN CNO availabilities	0.04	171,697
Puget Sound	Recycling	0.02	54,563
Puget Sound	Engineering	0.01	11,154

SOURCE: Norfolk Naval Shipyard, 2014; Pearl Harbor Naval Shipyard, 2014; Portsmouth Naval Shipyard, 2014; and Puget Sound Naval Shipyard, 2014.

The direct man-days executed at the shipyards include any inefficiencies resulting from material delays, rework, natural disasters, or other events affecting performance. To determine why the man-days increased, we must look to other sources. Next, we discuss in more detail the causes of increased direct man-days for SSBN, CVN, and other scheduled maintenance, as well as how these increases affect indirect workload.

Ballistic-Missile Submarine Maintenance

The SSBN fleet currently comprises *Ohio*-class vessels constructed between 1976 and 1993. These submarines are each 560 ft long, displace 18,750 tons while submerged, and are designed to carry 155 crew members. Fourteen SSBNs provide a nuclear-strike capability. As of 2008, all *Ohio*-class SSBNs carry D5 missiles. Four other converted *Ohio*-class vessels served as SSGNs supporting conventional-strike and other special-operations missions.

The SSBN uses a phased maintenance strategy to improve operational availability and readiness through more-frequent but shorter maintenance periods.[4] *Ohio*-class vessels are scheduled for three CNO availabilities over their 42-year service lives. The first and last availabilities are referred to as extended refit periods (ERPs), which the vessels spend in dry dock. Per OPNAV Notice 4700, the first ERP requires four months of work and occurs when a submarine has been in service for 14 years; the last ERP requires 5.2 months of work and occurs when a submarine has been in service for 33 years.[5] At mid-life, or after 21 years of service, the submarine enters engineered refueling overhaul (ERO), a major modernization and maintenance package that replaces the nuclear fuel core and performs other deep-maintenance tasks, such

---

[4] See Office of the Chief of Naval Operations, *Submarine Engineered Operating Cycle Program*, Instruction 3120.33C, January 22, 2013.

[5] OPNAV Notice 4700, *Representative Intervals, Durations, Maintenance Cycles, and Repair Man-Days for Depot Level Maintenance Availabilities of U.S. Navy Ships*, Office of the Chief of Naval Operations, U.S. Department of the Navy, 2013.

as pulling the shaft. EROs are planned for 27 months, with much of this time spent in dry dock for work that cannot be done pierside during continuous maintenance.

SSBNs also receive a series of short, more-frequent phased maintenance availabilities. These availabilities are performed at the operating base, using a mix of personnel from the shipyard, the ship's force, and contractors.

Figure 2.4 shows the stipulated Class Maintenance Plan requirements, in total man-days, for SSBNs in recent years. In FY 2004, the number of total man-days required to maintain the SSBN over the 42-year life cycle was slightly more than 4 million. This work included each of the three CNO availabilities. ERP work remained fairly consistent across years, while ERO requirements increased. Overall, we

Figure 2.4
Total Man-Days Required Across a Ballistic-Missile Submarine's Life Cycle, Selected Years

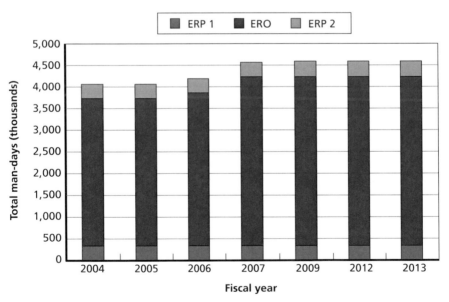

SOURCE: Class Maintenance Plan requirements from OPNAV Notice 4700, various years.
NOTE: No data were found for FYs 2008 and 2011.
RAND RR1552-2.4

observed a 13-percent increase in the planned CNO availability maintenance requirement from FYs 2004 through 2013.

Between FYs 2008 and 2014, the public shipyards conducted six EROs for SSBNs 731 through 736. Before FY 2011, the duration of an ERO was approximately 27 months, but afterward, the duration grew to more than 31 months, as shown in Figure 2.5. After FY 2010, as the duration of the availability increased, the number of direct man-days required to execute the availability also increased. At a minimum, the fixed costs associated with conducting the availability, including those for indirect man-days, increased proportionally with the duration of the ERO. The ERO requiring the most man-days, more than 550,000, was for SSBN-731 in FY 2008, while the ERO for SSBN-733 required a little more than 410,000 in FY 2010. The most-recent ERO required more than 540,000 direct man-days.

Figure 2.5
Duration and Man-Days Required for Completed Engineered Refueling Overhauls of Ballistic-Missile Submarines, Selected Years

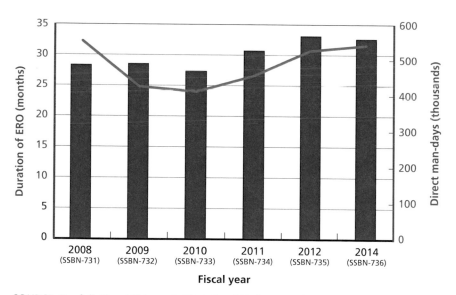

SOURCE: Norfolk Naval Shipyard, 2014; Pearl Harbor Naval Shipyard, 2014; Portsmouth Naval Shipyard, 2014; and Puget Sound Naval Shipyard, 2014.
RAND RR1552-2.5

To better understand the causes for the increases in duration and man-days required for SSBN maintenance, we reviewed shipyard and NAVSEA documentation that approved changes to the availabilities from FYs 2008 through 2014. These documents are referred to as *re-baseline letters*. The letters describe the initial request for resources, the approved re-baselining changes, and the terms and conditions of the changes. Four of the six availabilities in Figure 2.5 had an approved change during execution.[6]

The re-baseline letters described several causes for increases in workload, including new work exceeding planned levels, extensive unplanned repair growth, and resource constraints, all contributing to delays and increased requirements.[7] According to our discussions with those working most closely with the boats, resources are often diverted from the longer SSBN ERO to availabilities that are closer to completion or are higher priority. While this research was being conducted, Navy leaders changed the prioritization of work at the shipyards to mitigate this problem.

Nuclear-Powered Aircraft Carrier Maintenance

CVNs are the largest and some of the most-complex ships in the Navy. They are designed to house more than 5,000 sailors, embark 80 aircraft, and operate for 50 years. Around 1,100 ft long and displacing around 100,000 tons, they are also the centerpiece of the Navy's force structure, projecting force and presence around the world. The current oldest aircraft carrier in the Navy is the USS *Nimitz* (CVN-68), the lead ship of its class, commissioned in 1975. The newest aircraft carrier will be the USS *Gerald R. Ford* (CVN-78). It is the lead ship of its class, scheduled to be commissioned and placed in service in 2017. Figure 2.6 shows the USS *George Washington* (CVN-73) in maintenance.

---

[6]   During our research, we reviewed nearly two dozen re-baseline letters from the commanders of the naval shipyards. See the various change request letters in the bibliography of this report.

[7]   No additional documentation of the rationale is available.

Figure 2.6
***Nimitz*-Class USS *George Washington* (CVN-73) in Maintenance**

SOURCE: Peter D. Blair, U.S. Navy.
RAND *RR1552-2.6*

Since the mid-1990s, the CVN fleet has been on the Incremental Maintenance Plan (IMP). The IMP specifies depot maintenance packages at standard intervals and defines two primary maintenance packages. The first is a pierside planned incremental availability (PIA) that notionally lasts six months. The size of the PIA increases as the ship ages, from 146,200 direct man-days to 201,400 man-days. A PIA is a maintenance period during which certain machinery, including the aviation systems, can be repaired or replaced and other work can be done. The second availability type is the docking planned incremental availability (DPIA), which requires the ship to be dry-docked. DPIAs are larger than PIAs, lasting at least 16 months and requiring more than 500,000 direct man-days of work. Maintenance activities that require a longer period of time, such as installing new communication systems and pulling and replacing shafts and rudders, are completed during a DPIA. A post-shakedown availability (PSA) is performed after the ship is first delivered to the fleet and after the refueling and complex overhaul (RCOH). These availabilities provide opportunities to address any outstanding readiness issues prior to deployment.[8]

Currently, the IMP calls for maintenance to occur on a PIA-PIA-DPIA cycle, which repeats through the life of the carrier, interrupted only by the mid-life RCOH. Figure 2.7 shows a notional carrier life in terms of these maintenance availabilities.

Figure 2.7
Notional Nuclear-Powered Aircraft Carrier Life-Cycle Maintenance Plan

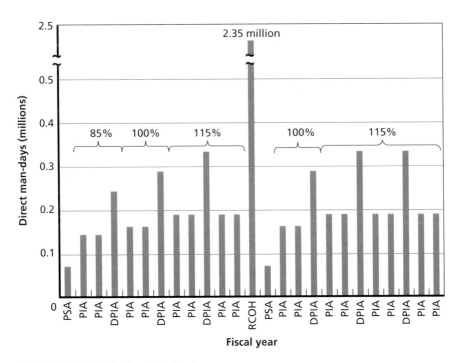

SOURCE: OPNAV Notice 4700, 2004.
NOTE: Each PIA-PIA-DPIA cycle of availabilities receives a proportion of the man-days expected for the next cycle of availabilities, on average. Percentages shown indicate these proportions.
RAND RR1552-2.7

---

8   OPNAV Notice 4700, 2004.

In the 2004 Class Maintenance Plan, a carrier received 16 PIAs and six DPIAs over its expected life cycle. At the beginning of a CVN life cycle, the availabilities receive 85 percent of the expected man-days required in the next set of availabilities and 115 percent prior to the RCOH, as shown in Figure 2.7. This is done to adjust the planned work for the aging of the ship. Early availabilities, historically, have not required as much work as the notional availability plan states, while later availabilities, when the ship is older, require more work than the notional schedule projects. The RCOH is the largest of the aircraft-carrier maintenance packages, and each RCOH is performed by Newport News Shipbuilding rather than in the naval shipyards. The RCOH maintenance package requires more than four years to complete. During an RCOH, the nuclear reactors are refueled; all machinery is updated, repaired, or replaced; major electrical work may be performed; and propellers, shafts, and rudders are refurbished and replaced, among other tasks. The RCOH essentially "resets" the age of the ship before its second half of life. In that second half, the PIA-PIA-DPIA cycle repeats.

The aircraft-carrier life-cycle maintenance requirement as described in OPNAV Notice 4700 has fluctuated by 1.3 million total man-days between the highest planned life-cycle maintenance (FY 2004) and the lowest (FY 2006), shown in Figure 2.8.

The Optimized Fleet Response Plan (OFRP) has been designed to result in more time between maintenance periods to increase the time that a carrier is available for tasking. Additionally, it is intended to align the operational and maintenance schedules of all the ships in the carrier strike group. The OFRP life-cycle man-day requirements that were expected in FY 2015 are very similar to those that were required to maintain aircraft carriers nearly 20 years ago. At that time, 25 CNO availabilities were planned for each aircraft carrier over the life of each ship. In the OFRP in FY 2015, there are 16 CNO availabilities planned, and a carrier requires nearly 7.5 million total man-days over its life. Ten years ago, these CNO maintenance activities were completed over 25 CNO availabilities, resulting in 300,000 man-days per availability, on average. This same requirement completed over just 16 availabilities increases the work per CNO availability to 470,000, on average. How-

Figure 2.8
Total Maintenance Man-Days per Nuclear-Powered Aircraft Carrier's Life Cycle, Selected Years

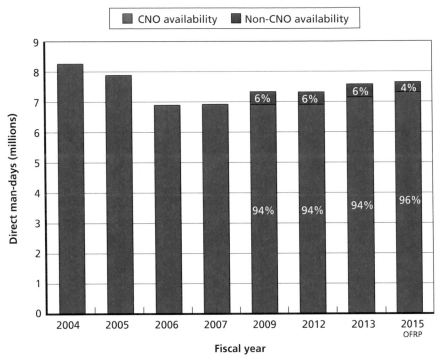

SOURCE: OPNAV Notice 4700, various years.
NOTE: No data were found for FYs 2008, 2011, and 2014.
RAND *RR1552-2.8*

ever, there are limits to the amount of man-days that can be performed per day at each shipyard. To help alleviate this additional demand, non-CNO availabilities have been added to the life-cycle maintenance plan. These are known as continuous incremental availabilities and are notionally planned for the OFRP to have 45 days duration, require 28,000 man-days to execute, and be accomplished at the operating station. A continuous incremental availability is planned for each carrier between non-docking CNO availabilities. At present, 12 such availabilities are notionally scheduled during the full life of a single carrier.

While adding more continuous maintenance has helped provide opportunities to complete some work that no longer fits within the

CNO availabilities, the PIA and DPIA have also increased in size, as Figure 2.9 shows.

Between FYs 1992 and 2015, the interval between each successive maintenance-operational and training cycle has changed from 18 months to 24 months to 27 months to 32 months to the planned OFRP plan of 36 months.[9] These changes have resulted in a carrier being available for tasking for longer periods between maintenance, but the improved availability has implications for maintenance. In particular, the changes to the operational cycle have resulted in larger and

Figure 2.9
Direct Man-Days Executed for PIAs and DPIAs, Selected Years

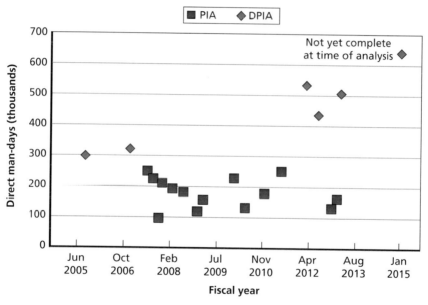

SOURCE: Carrier Planning Activity, "Deployment History," spreadsheet, U.S. Department of the Navy, provided to RAND by Carrier Planning Activity, February 10, 2014a.
NOTE: No data on PIA or DPIA executions were delivered for FYs 2007, 2011, or 2014.
RAND RR1552-2.9

---

[9] For further information on operational cycles and impacts of changing the cycle, see Roland J. Yardley, John F. Schank, James G. Kallimani, Raj Raman, and Clifford A. Grammich, *A Methodology for Estimating the Effect of Aircraft Carrier Operational Cycles on the Maintenance Industrial Base*, Santa Monica, Calif.: RAND Corporation, TR-480-NAVY, 2007.

longer DPIAs. Each successive lengthening of the CVN's operational cycle resulted in maintenance being rephased as CNO availabilities were reduced in number and expanded in scope. In the most recent change, a PIA was eliminated from the maintenance plan, and the work was absorbed, at least partially, by the remaining DPIAs. At the same time, carrier deployment lengths have grown. The DPIAs performed in FYs 2012 and 2013 were completed after much longer deployments than initially planned, and there was additional work performed in these availabilities, particularly in shafting and rudders. The Navy's Carrier Planning Activity assesses that this additional work was driven partially by operational needs, reflecting longer duration between dry dockings, and partially by the aging of the carrier force.[10]

Although the overall carrier fleet size has decreased, the number of CVNs being maintained in the naval shipyards will have increased by 38 percent between FYs 2001 and 2016. In FY 2001, the shipyards were supporting eight CVNs. When CVN-78 is commissioned in FY 2017, the Navy will operate, and the public shipyards will support, 11 CVNs. This seemingly small increase in the number of carriers that the shipyards must support equates to a significant increase in shipyard maintenance because only two shipyards that do carrier maintenance remain—Norfolk and Puget Sound. Over its life cycle, an aircraft carrier requires nearly 7.5 million total man-days of maintenance—or about 150,000 man-days per year. In other words, adding three carriers to the fleet will result in an average annual increase of approximately 450,000 man-days of maintenance.

Scheduled Maintenance

Scheduled maintenance is a term primarily used to describe work not accomplished in a CNO-level availability. CNO-level availabilities are the larger, more-complex availabilities during which most of the life-cycle maintenance and modernization are executed. These include aircraft carrier PIAs and DPIAs, SSN engineered overhauls and EROs,

---

[10] Based on conversations with the Carrier Planning Activity.

and SSBN EROs. Examples of scheduled maintenance are continuous maintenance activities, ship alterations, nuclear equipment disposal, fleet maintenance availabilities, Nuclear Regional Maintenance
Department activities, fleet technical support, availability planning
activities, and process activities. Although none of these activities
individually constitutes a significant portion of the public-shipyard
workload, together, they account for nearly one-fifth of the total direct
workload. Figure 2.10 shows the direct man-days for scheduled maintenance executed and planned at the shipyards through FY 2020.

Scheduled maintenance has increased significantly in recent years,
from 11 percent of direct man-days executed between FYs 2004 and
2008 to 16 percent between FYs 2009 and 2014. Detailed analysis of
each shipyard's scheduled maintenance workload indicates that these
increases occurred for various reasons. For example, at Norfolk Naval

Figure 2.10
Direct Man-Days of Scheduled Maintenance Executed and Planned,
FYs 2004–2020

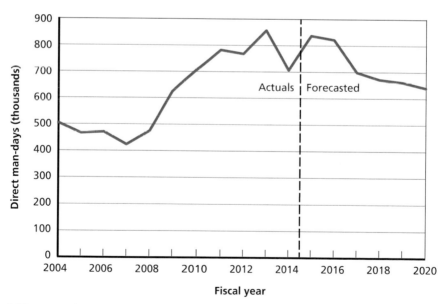

SOURCE: Norfolk Naval Shipyard, 2014; Pearl Harbor Naval Shipyard, 2014;
Portsmouth Naval Shipyard, 2014; and Puget Sound Naval Shipyard, 2014.
RAND RR1552-2.10

Shipyard, increasing intermediate-level surface ship work has been a key driver in scheduled maintenance, and although such maintenance had a sharp drop in FY 2014 for budgetary reasons, it is expected to recover in FY 2015 and beyond. At the other shipyards, CNO workloads have driven increases. Figure 2.11 shows the distribution of work within scheduled maintenance by shipyard and platform type.

Scheduled maintenance workloads for SSBNs and surface ships have increased from FYs 2008 through 2014, as shown in the figure. For CVNs and SSNs, most work has been conducted on fleet maintenance availabilities and continuous maintenance availabilities. The increases observed here may be evidence of the movement of maintenance activities from CNO availabilities to other scheduled maintenance periods within the operating cycle.

As seen in Figure 2.10, each of the four shipyards' scheduled maintenance increased between FYs 2008 and 2014. When trend lines are plotted for the direct man-days of scheduled maintenance at each yard individually, the increase is between 31 and 198 man-days per month (see Table 2.2).

Table 2.2
Average Increase per Month of Direct Man-Days of Scheduled Maintenance, FYs 2008–2014

Naval Shipyard	Trend Line Slope (direct man-day increase per month)
Norfolk	197.77
Portsmouth	77.14
Pearl Harbor	67.53
Puget Sound	31.24

SOURCE: Norfolk Naval Shipyard, 2014; Pearl Harbor Naval Shipyard, 2014; Portsmouth Naval Shipyard, 2014; and Puget Sound Naval Shipyard, 2014.

Figure 2.11
Scheduled Maintenance Direct Man-Days, by Shipyard and Platform Type, FYs 2008–2014

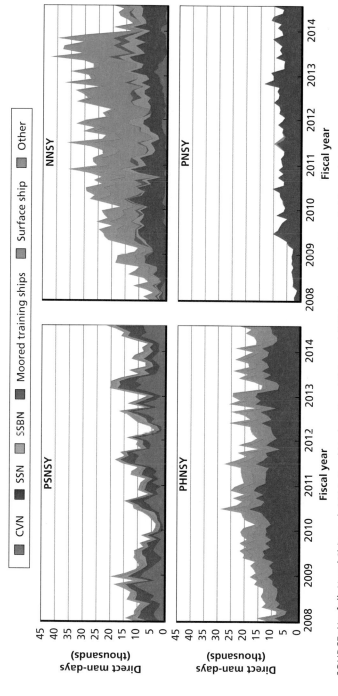

SOURCE: Norfolk Naval Shipyard, 2014; Pearl Harbor Naval Shipyard, 2014; Portsmouth Naval Shipyard, 2014; and Puget Sound Naval Shipyard, 2014.

RAND *RR1552-2.11*

Indirect Man-Days

NAVSEA defines *indirect costs* as those that cannot be directly linked to a final product or service. There are two basic types of indirect costs. *Productive costs* are those identified with a production cost center, including, for example, clerical and administrative support, training, and set-up time. *General and administrative overhead costs* are those that benefit the entire activity, including, for example, the office of the comptroller. The hours worked by personnel performing functions categorized as overhead are charged to indirect labor. NAVSEA is working to improve the consistency of accounting across the shipyards, but there is still some variability in how the yards account for indirect costs, including leave, training activities, supervisory tasks, environmental and safety activities, and process activities.

Indirect man-days have increased from about 1.5 million in FY 2008 to nearly 2.5 million in FY 2014, as shown in Figure 2.12. When compared with the increase observed in direct man-days (shown in Figure 1.1), the proportion of indirect man-days to direct man-days has increased. In FY 2008, the proportion of indirect to direct was about 0.625 indirect man-days for every direct man-day. In FY 2014, the ratio executed was 0.83 indirect man-days for every direct man-day.

Indirect man-days have increased for several reasons. Above all, as noted, indirect man-days increase as a function of direct man-days, which have increased. NAVSEA collects data on indirect charges, but it is difficult to assess the impact of new regulatory and headquarters requirements, which affect the indirect and direct man-hours required. These costs often show up in charges as part of other work. For example, if it now takes two hours to secure a compartment of a ship because of a new safety requirement, the record would show that it takes two hours to secure the ship. The record would not indicate that this activity used to take one hour before the new requirement. This makes it difficult to understand how much of an effect additional safety requirements have on indirect labor. Similarly, the savings resulting from investments to improve efficiencies are also difficult to discern. In addition to increases

Figure 2.12
Indirect Man-Days Executed at Public Shipyards, FYs 2008–2014

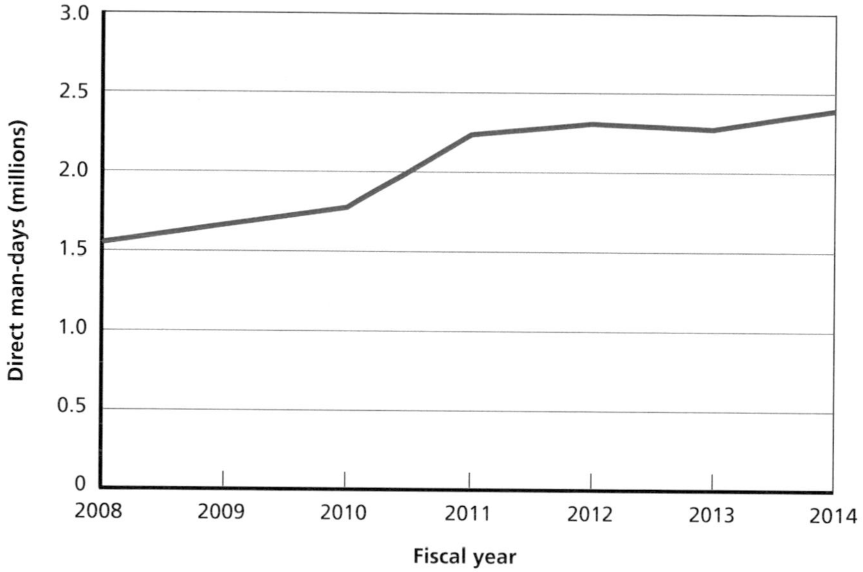

SOURCE: Norfolk Naval Shipyard, 2014; Pearl Harbor Naval Shipyard, 2014;
Portsmouth Naval Shipyard, 2014; and Puget Sound Naval Shipyard, 2014.
RAND RR1552-2.12

in safety requirements,[11] possible reasons for the rise in indirect workload include increased training and administration, more regulatory
occupational safety and health requirements for fall protection, facility maintenance and janitorial maintenance costs that were formerly
covered elsewhere, a NAVSEA policy to shred all project paper with
writing, increases to workforce development improvement and investments, increased environmental and regulatory compliance requirements, and implementation of processes and technology to improve
auditability. In addition, the inclusion of the Regional Maintenance

[11] Shipyard representatives gave the example of increased safety requirements in the NAVSEA
Industrial Ship Safety Manual for Fire Prevention and Response, which were a response to a
fire on the USS *Miami* (SSN-755). Representatives noted that the policy changes increased
indirect man-days, but tracking of the new requirements had not yet begun.

Centers in the NAVSEA management portfolio could have also contributed to the increase in indirect man-days.

The shipyards' WARRs provide some information about how much time is spent performing indirect activities. We observe that most indirect expenses are incurred within four primary categories: leave, training, supervisory, and overhead (a residual category). Figure 2.13 shows the percentage of the workforce in each category by month from FYs 2004 through 2016.

Training has increased slightly, while leave and supervisory percentages have remained fairly consistent. The overhead category has been consistently around 20 percent since FY 2006 but was slightly smaller, around 15 percent, before then.

Figure 2.13
Percentage of Workforce Charging to Leave, Training, Supervisory, or Overhead, FYs 2004–2016

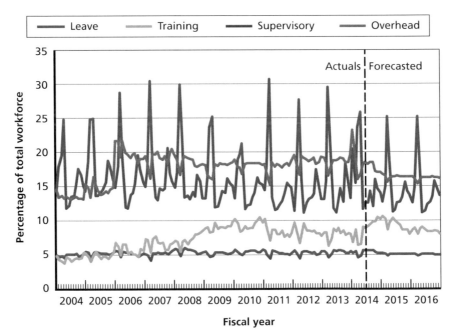

SOURCE: Norfolk Naval Shipyard, 2014; Pearl Harbor Naval Shipyard, 2014; Portsmouth Naval Shipyard, 2014; and Puget Sound Naval Shipyard, 2014.
RAND RR1552-2.13

In our discussions with shipyard representatives and in reviewing the growth in engineering personnel at the shipyards, we found that more effort in engineering could also contribute to some of the increases in indirect labor, but data were not available to evaluate this hypothesis. In the case of engineer effort required per inducted direct man-day or per technical work document, there was a strong sense, and many in our discussions expressed, that the required engineering work has increased. Examination of both nuclear and non-nuclear personnel at the shipyards between FYs 2008 and 2014 shows increases in manning.

Summary

The direct and indirect man-days executed at public shipyards have increased for a variety of reasons. For instance, the amount of work the shipyards must accomplish has increased (as shown in Table 2.1). This additional work is a result of new life-cycle maintenance work being discovered (which is later incorporated into the baseline life-cycle maintenance plans), additional unplanned work, and increases to the number of platforms the shipyards must support. Furthermore, some increases result from constrained resources and shipyard performance. Increases to indirect labor are greater than the increases observed in direct labor, partly because of increased training and more regulatory and policy requirements, but additional information must be collected to better explain these increases.

CHAPTER THREE

Are More Personnel Now Required to Perform the Same Work?

For ships serviced at the public shipyards, the maintenance requirement is initially developed independently from what it takes to execute the work. In other words, the requirement is expressed as an activity that needs to be performed (e.g., a pump must be changed or a tank must be inspected). The level of effort required to perform this work will depend on the availability of skilled labor, the accessibility of equipment, and the condition of the equipment, among other variables.

In this chapter, we focus on estimating the effects of an increasingly inexperienced workforce. Specifically, we estimate the man-days that can be generated by the workforce as levels of experience change.[1] This is not because the other contributing factors are less important but because there is little data to help us to define their effects. Here, we analyze workforce data to better understand the impact of an increasingly inexperienced workforce on the amount of man-days required to execute a given requirement.

Workforce Demographics

The shipyards employ civilians, military personnel, and contractors to execute their work. Theoretically, if the shipyards maintain a steady or increased amount of work, a reduction in military personnel or con-

---

[1] Using other measures of output—such as pump overhauls, piping replacements, and breaker repairs—is another approach to estimating the productivity of the shipyards' workforce.

tractor support would imply a need to increase the number of civilians. We do not observe a decrease in the number of military personnel, according to the Navy's budget exhibits.[2] In addition, conversations with shipyard management have indicated that contract support has either remained steady or increased, depending on the shipyard in question. This allows us to assess trends with the civilian workforce.

In the early 1990s, the public shipyards saw a dramatic decline in the civilian workforce. There was little hiring during this time. This created the commonly-referred-to "bathtub effect," whereby an organization has many staff with little experience (as defined by years on the job), many staff with much experience, and few staff between these levels. This poses particular challenges for the public shipyards when they need to grow capability because they cannot easily hire individuals with the required skills and experience from the private sector. Figure 3.1 shows the distribution of the workforce by years of experience in FYs 2006 through 2014. The bathtub effect is evident in FY 2006, when 35 percent of the civilian workforce across all four shipyards had less than ten years of service, and 50 percent had at least 20 years of service. In contrast, in FY 2014, 48 percent of the workforce had less than ten years of experience, and only 28 percent had at least 20 years of experience. The workforce has gotten less experienced.

While the distribution of the experience is distinctly different between FYs 2006 and 2014, the proportion of wage grade personnel to general schedule personnel has remained fairly constant during that span. Figure 3.2 shows the increase in civilian staffing for each personnel category and the proportion of workers who are general schedule. Wage grade personnel are typically associated with a shipyard's wrench-turning activities that are considered direct labor, and general schedule employees are typically associated with management or engineering activities that are considered indirect or overhead activities.[3]

---

[2]   These data represent the authorized billets. The number of billets filled may be less than the number authorized. See U.S. Department of the Navy, various years.

[3]   There are general schedule employees who charge to direct labor categories and wage grade employees who charge to indirect labor categories.

Figure 3.1
Civilian Workforce Experience, FYs 2006–2014

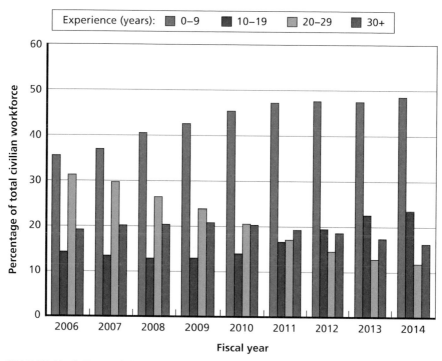

SOURCE: Norfolk Naval Shipyard, 2014; Pearl Harbor Naval Shipyard, 2014;
Portsmouth Naval Shipyard, 2014; and Puget Sound Naval Shipyard, 2014.
RAND RR1552-3.1

Figure 3.2
Wage Grade and General Schedule Civilian Personnel, FYs 2006–2014

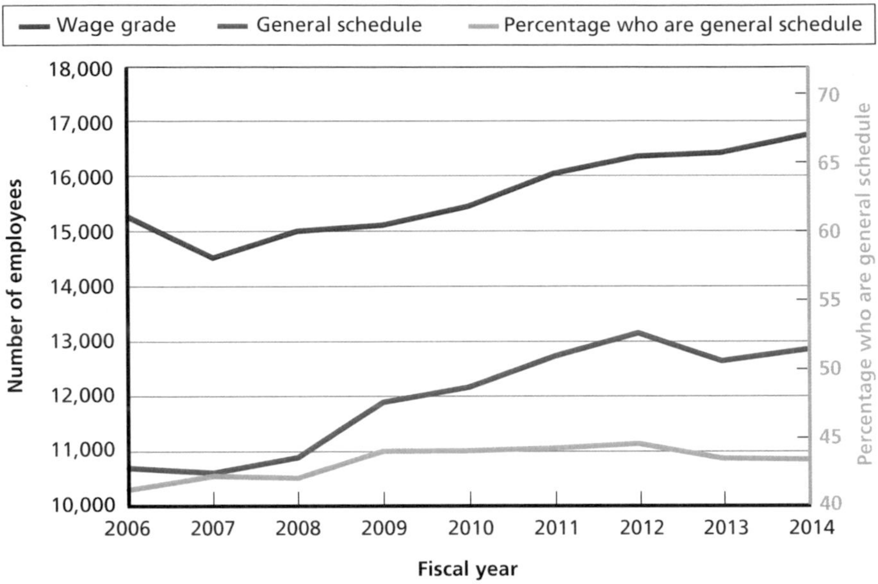

SOURCE: Norfolk Naval Shipyard, 2014; Pearl Harbor Naval Shipyard, 2014;
Portsmouth Naval Shipyard, 2014; and Puget Sound Naval Shipyard, 2014.
RAND RR1552-3.2

Relative Productivity

Productivity can be defined in different ways. It can be the relative
fraction of productive work over a unit of time, or it may be the breadth
of tasks a worker is capable of accomplishing.[4] Here, we define *productivity* as the relative fraction of productive work over a unit of time.
Expert elicitation was used to define the relative output of an individual with experience compared with the output of an individual with
less experience. RAND researchers have performed several studies that
examined the relative productivity of "green"—that is, inexperienced—
labor with experienced labor in the production trades at nuclear ship-

---

[4]   Specific measures of shipyard or shop-level output generated by the workforce could also
be used to define productivity.

yards. In a 2007 report, Schank and colleagues found that it takes approximately eight years for a new worker to become fully productive at a nuclear-vessel shipyard.[5] In 2008, Riposo and colleagues found that it takes seven years for a new worker to become fully productive in the public shipyards.[6] In particular, Riposo and colleagues found that apprentices with 2.5 years of experience are half as productive as a journeyman with seven years of experience, and a journeyman with five years of experience is 90 percent as productive as a journeyman with seven years of experience. Some individuals, and some trades, take more or less time to achieve these levels of productivity.

There are new programs and efforts under way at the shipyards to help expedite the amount of time it takes a production shop trades-person to become fully proficient. Trade superintendents at the shipyards indicated that it takes an individual about seven years to become fully proficient. However, a new training program, which provides an average of two months of "trade skill boot camps," has improved the initial productivity since its inception in FY 2007. We refer to the new training program throughout this report as *accelerated training*. Accelerated training initiatives that have been implemented for high-solids paint systems in submarine ballast tanks successfully yielded work execution by trade employees with half of the experience time at the shipyard, including improved quality of work.

Figure 3.3 shows the relative productivity of workers in production trades with different years of experience, both historically and for those who have undergone accelerated training. An individual with one year of experience can produce, on average, about one-quarter of what a fully experienced individual with seven years of experience can produce. Increases are generally linear after that, with the accelerated

[5] John F. Schank, Mark V. Arena, Paul DeLuca, Jessie Riposo, Kimberly Curry, Todd Weeks, and James Chiesa, *Sustaining U.S. Nuclear Submarine Design Capabilities*, Santa Monica, Calif.: RAND Corporation, MG-608-NAVY, 2007.

[6] Jessie Riposo, Brien Alkire, John F. Schank, Mark V. Arena, James G. Kallimani, Irv Blickstein, Kimberly Curry Hall, and Clifford A. Grammich, *U.S. Navy Shipyards: An Evaluation of Workload- and Workforce-Management Practices*, Santa Monica, Calif.: RAND Corporation, MG-751-NAVY, 2008.

Figure 3.3
Average Relative Productivity for Years of Service, Historically Versus Accelerated Training

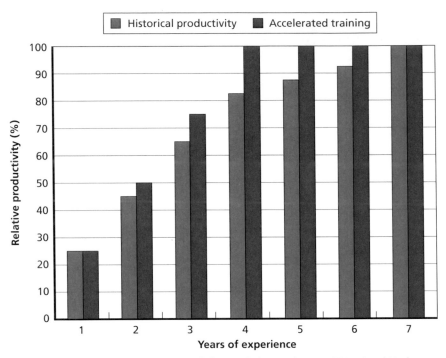

SOURCE: RAND analysis based on Norfolk Naval Shipyard, 2007, 2014; Pearl Harbor Naval Shipyard, 2007, 2014; Portsmouth Naval Shipyard, 2007, 2014; and Puget Sound Naval Shipyard, 2007, 2014.
NOTE: The historical productivity numbers are from FY 2007. The accelerated training numbers are an average from FYs 2008 through 2014.
RAND RR1552-3.3

training program having some success in reducing the time required to make a worker fully proficient.

Ideally, there would be a steady flow of new hires, hired far enough in advance to develop expertise and to replace experienced labor over time. The public shipyards, however, have experienced many hiring freezes and layoffs that have resulted in a workforce that is not distributed normally. Most recently, the Budget Control Act of 2011 caused the shipyards to curtail hiring because of budget reductions.[7] The proportion of the workforce that is experienced is historically small, while the proportion of the workforce that is inexperienced is historically large. The result is experienced workers being replaced by less-experienced workers. New employees are hired at low wage grades. Workers at the lowest grade have a lower level of productivity but also a lower hourly wage than a more experienced worker. As the newly hired workers replace the more-experienced workers, the hourly wage rates decline, but productive work also declines. Previous work has shown that the cost of a fully productive man-day varies.[8] An apprentice with 2.5 years of experience is half as productive and costs 30 percent less than a worker with seven years of experience. However, a new hire is one-fourth as productive and costs 40 percent less. The remainder of this chapter assesses how significant the impact to productivity from this less-experienced workforce will be and when NAVSEA might see improvement.

Estimating the Future Workforce

If we look at the relative productivity of the civilian workforce in FY 2007 and compare that with the relative productivity of the workforce in subsequent years, we can observe a relative change in the overall proficiency of the workforce. At any point in time, the relative productivity of the workforce is equal to the sum of the number of individuals in each experience level (e_i) multiplied by the relative productivity (p_i) of each experience level. Or, mathematically,

$$\text{Productivity} = \sum_{i=1}^{n}(e_i \times p_i).$$

[7]   Public Law 112-25, Budget Control Act of 2011, August 2, 2011; Deborah McDermott, "Sequestration Threatens Portsmouth Naval Shipyard Jobs, Workload," *Seacoast Online*, September 20, 2012.

[8]   Riposo et al., 2008.

Using data provided by NAVSEA 04, we used the number of individuals within each level of experience and the historical productivity factors identified in Figure 3.2 to estimate an overall productivity level for the workforce for each year from FYs 2007 through 2014.[9]

Table 3.1 shows the productivity of the civilian workforce by year relative to that in FY 2007. The reduction in productivity since FY 2007 can manifest itself as an increase in unplanned overtime, additional straight-time man-days, or a need to outsource work. All of these activities increase cost.

To understand how the relative productivity of the workforce will be affected beyond FY 2014, we must forecast the future workforce. To do this, we apply the same methodology employed by Schank and colleagues.[10] We begin with the number of workers in established five-year age brackets (e.g., ages 21 to 25) and experience level (defined as years of service). In each time-step of one fiscal year, some workers retire, some attrite for other reasons, some move to the next age bracket, and

Table 3.1
Productivity of the Civilian Workforce
Relative to FY 2007, FYs 2007–2014

Fiscal Year	Relative Productivity (%)
2007	100.0
2008	98.0
2009	97.4
2010	96.4
2011	95.2
2012	95.3
2013	96.7
2014	96.1

[9] The model was verified through use of a training set.

[10] Schank et al., 2008, pp. 147–148.

all gain an additional year of experience. Individuals have a productivity value depending on their years of experience.

For each year, we seek to identify the following:

- the "effective workforce," which is the number of fully productive employees
- the "predicted need," which is the number of individuals we estimate the shipyard needs to hire to achieve that productivity level
- the "planned workforce," which is what NAVSEA has specified it requires to execute the estimated workload.

Hiring occurs to the maximum level possible, which is 25 percent, or to a level that allows for the productive workforce to equal the planned level.[11]

To estimate the future workforce, we need to make a variety of assumptions. Many of these are based on previous research findings or historical data. The most important assumption here is that it is possible to make up lost productivity through hiring. In other words, we assume that if a fully productive individual requires four hours to accomplish a work package, two individuals who are half as productive can accomplish the same work in four hours. The other assumptions include the following:

- The shipyards can hire up to 25 percent of the current civilian workforce in a year.
- Changes to productivity are relative to FY 2007.
- Lost productivity can be recovered through hiring.
- Overall attrition is 5 percent.
- All workers retire when reaching age 66.
- All new hires are inexperienced.
- "Planned need" refers to NAVSEA's plan to increase the workforce to 33,500 by FY 2017.
- Of the civilian workforce, 55 percent, or 18,425, work in the production trades.

---

[11] Industry subject-matter experts from both public and private nuclear shipyards suggested in interviews that growth of the workforce is possible up to a maximum of 25 percent of the total workforce.

- On average, it takes seven years to become fully productive using traditional training approaches.
- On average, it takes four years to become fully productive using accelerated training approaches.

The appendix to this report provides sensitivity analyses. Table 3.2 shows predicted productivity, relative to FY 2007, of our model. Our predicted results are very similar, though not identical, to actual pro-

Table 3.2
Model-Predicted Productivity of the Civilian Workforce Relative to FY 2007, FYs 2007–2023

Fiscal Year	Relative Productivity (%)
2007	100.0
2008	98.3
2009	97.7
2010	96.8
2011	95.7
2012	96.1
2013	97.7
2014	97.2
2015	92.8
2016	89.3
2017	87.6
2018	90.8
2019	93.4
2020	94.3
2021	94.8
2022	95.5
2023	95.6

ductivity observed from FYs 2008 through 2014. More importantly, our model predicts a decrease in productivity through FY 2018 and then a gradual increase in productivity as the workforce reaches a steady state of hiring and new hires accumulate experience. From FYs 2015 through 2018, the shipyards are hiring to reach their planned civilian staffing levels. Beyond FY 2019, once the hiring goals are met, hiring occurs only to offset attrition.

This forecasted dip in productivity in the coming years is particularly problematic because there is a peak in workload during this time, which is the impetus to the hiring. The Navy has some options to address this misalignment, such as additional hiring and outsourcing work. Because of the magnitude of the loss of productivity and the constraints on hiring (no more than 25 percent of the total workforce can be hired in any given year), it is infeasible to recover productivity entirely through additional hiring. The shipyards would have had to hire tens of thousands of workers in FY 2015 alone, which was impractical and was not accomplished. The shipyards are indeed hiring to reach the planned 33,500 workforce target, but not at the rate shown in the research modeling. As a result, NAVSEA has already begun to outsource some work for coming years.

Figure 3.4 shows the model-predicted need, resulting effective workforce levels, and planned workforce levels for coming years. The ultimate planned workforce for the production trades is 18,425. To achieve an effective workforce of this size, NAVSEA will need to increase the production workforce to 25,000. Our effort indicated that in FY 2015, NAVSEA would need to hire an additional 2,000 workers. In reality, the shipyard workforce grew by nearly 3,000 employees in FY 2015, with around 750 more added in FY 2016 through mid-August 2016.[12] With these additional hires, the effective workforce will equal the planned workforce by FY 2017. After that, the effective workforce will continue to increase as the new hires accumulate experience and gain productivity. The effective workforce will exceed the currently identified planned workforce, but it can be reduced by hiring at a rate that does not fully replace attrition, as we do in our estimation.

---

[12] NAVSEA 04, meeting with the authors, August 19, 2016.

Figure 3.4
Predicted, Effective, and Planned Production Workforce, Model Simulation,
FYs 2015–2023

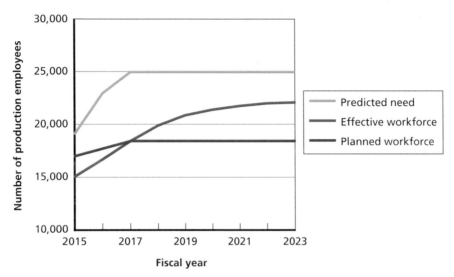

SOURCE: RAND analysis based on Norfolk Naval Shipyard, 2007, 2014; Pearl Harbor
Naval Shipyard, 2007, 2014; Portsmouth Naval Shipyard, 2007, 2014; and Puget Sound
Naval Shipyard, 2007, 2014.
RAND RR1552-3.4

If accelerated training is successful, the effective workforce will
reach the planned level a few months earlier. More importantly, such
training reduces the predicted need for labor, as shown in Figure 3.5.
Under the historical productivity levels, nearly 25,000 production
employees were needed; under the new training model, the predicted
need is only 23,000.

Summary

While there are noted benefits to having greater agility and familiarity
with modern technologies, it is just as important to develop expertise
and become fully productive in a trade or craft, which takes time.[13]

---

[13] Discussions with naval shipyard staff.

**Figure 3.5
Predicted, Effective, and Planned Production Workforce, Traditional Versus
Accelerated Training Programs, FYs 2015–2018**

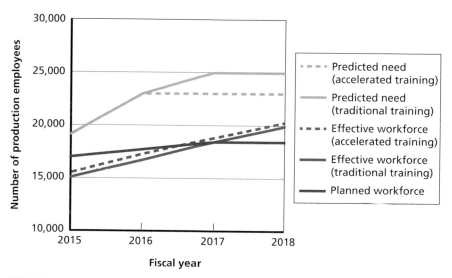

SOURCE: RAND analysis based on Norfolk Naval Shipyard, 2007, 2014; Pearl Harbor
Naval Shipyard, 2007, 2014; Portsmouth Naval Shipyard, 2007, 2014; and Puget Sound
Naval Shipyard, 2007, 2014.
RAND RR1552-3.5

Accelerated training programs may reduce the amount of time needed
to reach full productivity in certain trades. Regardless, our productivity model predicts that the effective workforce will not be sufficient
to accomplish the entire naval shipyard workload planned through
FY 2018. The predicted need may result in excess labor in the middle
term, but such additional workers could help the Navy address risks of
still-greater workload. Alternatively, changes to future hiring rates can
reduce the civilian workforce level to a more desirable number while
allowing the Navy to better meet near-term goals.

CHAPTER FOUR

Future Risks and Challenges

There is inherent uncertainty in naval shipyard work. Some of the maintenance activities are planned, and many are emergent. And although some maintenance activities are planned to require a certain number of man-hours and materiel, there is variability in execution. In addition, there are many nontechnical contributors to the uncertainty in future workloads. The requirements are based on assumptions about how the platforms operate, but operational needs can change without notice or consideration of impacts to maintenance. The acquisition, budgeting, and availability-planning processes can increase the uncertainty. Material assessments for condition-based maintenance are typically only finalized within an 18-month period ahead of major availabilities, during which the condition of the ship may change.

In this chapter, we seek to identify additional challenges that may lead to further risk and requirement increases in future planning.

Historical Trends

One of the potential risks to the future budget plan is the historical precedence of increasing funding requirements as execution nears. Table 4.1 shows the increases in shipyard resources from the initial budget submission to execution year, as outlined in the Navy's budget exhibits. The left-most column represents the budget year. The second column indicates what was initially proposed for that year (as shown in the President's budget request for that year). The third column shows what was enacted for that year, a value that may change because of

Table 4.1
Budgeted, Enacted, and Actual Expenditures for Shipyard Resources,
FYs 2008–2014

Fiscal Year	Budgeted ($ millions)	Enacted ($ millions)	Actual ($ millions)	Percentage Increase from Budgeted to Actual
2008	3,273	3,437	3,877	18.4
2009	3,364	3,824	4,015	19.3
2010	3,718	3,750	4,359	17.2
2011	3,973	3,877	4,649	17.0
2012	3,842	4,027	4,409	14.8
2013	4,031	4,031	4,304	6.8
2014	4,126	3,860	4,653	12.8

SOURCE: U.S. Department of the Navy budget exhibits (see U.S. Department of the Navy, various years).

reprogramming, rescissions, and other program alterations. The fourth column shows what was ultimately spent in that year, and the fifth column shows the percentage increase from budgeted to actual.

The estimate for FY 2008, submitted in the FY 2008 President's budget, was for $3.27 billion. In the FY 2009 President's budget submission, the enacted amount for FY 2008 was $3.44 billion. In the FY 2010 submission, the actual expenditures were $3.88 billion. In other words, for FY 2008, the Navy required more than half a billion dollars above what it requested in the FY 2008 President's budget for FY 2008.

From FYs 2007 through 2014, there has been a consistent trend of increases in required funding for naval shipyards from budget submission through execution. The average increase over this period is approximately 13 percent. There are, no doubt, complex reasons for this trend, including ship operations, current maintenance execution, dynamic shipyard performance for work under way in the planning years, increases in required maintenance identified late in the planning process, and increases resulting from delayed delivery of ongoing maintenance. Nevertheless, the historical trend for current planning and budgeting methodology consistently underestimates required funding levels. If changes are not made to curtail this trend, the Navy should expect to see similar increases in future budget submissions. For example, a 13-percent increase to the $4.25 billion proposed for FY 2016 would result in an actual requirement of approximately $4.8 billion.

Figure 4.1 shows the actual and projected budget trends from the current corporate, long-range naval shipyard WARRs for future years, assuming the same increases as observed historically. The area with diagonal lines represents the additional scope of work required if historical budget trends continue and there is a 13-percent increase

Figure 4.1
Direct Man-Days Executed and Planned at Public Shipyards with Additional Workload to Correct for Likely Budget Underestimates, FYs 2010–2021

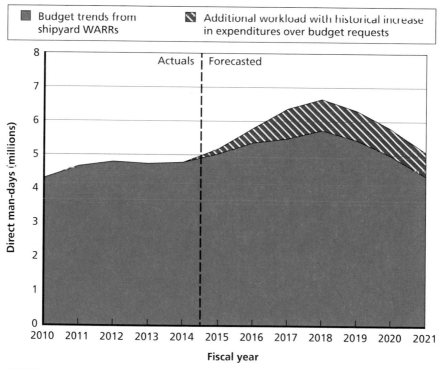

SOURCE: RAND analysis based on Norfolk Naval Shipyard, 2014; Pearl Harbor Naval Shipyard, 2014; Portsmouth Naval Shipyard, 2014; and Puget Sound Naval Shipyard, 2014.

RAND RR1552-4.1

in expenditures over budget requests. This added workload would increase the already high peak in workload expected for FY 2018. It would also offset, to some degree, the predicted workload decrease in the early 2020s when the SSBN ERO availabilities and SSN-688–class engineered overhaul availabilities are completed.

The Aging of Nuclear Assets

An additional future risk factor that could have significant consequences both for the Navy operationally and for maintenance execution in the naval shipyards is the aging trend that will occur in the near and middle terms for ship classes maintained by the shipyards. Figure 4.2 depicts the percentage of nuclear submarines with more than 25 years of service and the percentage of nuclear aircraft carriers with more than 40 years of service from FYs 1980 through 2040. We chose these thresholds because they represent the point when each ship is ten years from planned retirement. Experience with such ships as the USS *Enterprise* (CVN-65), USS *Nimitz* (CVN-68), and some of the *Los Angeles*–class submarines has shown that a ship requires large maintenance increases in the final ten years of its life cycle.

    Ohio-class SSBNs will all reach at least 25 years of service life by 2020, with the class aging out of service in the subsequent decade. And in the coming years, most of the SSN fleet will have at least 25 years of service life. Although this proportion will later diminish, the proportion of SSNs in later life for several years will be at unprecedented levels and will pose risks for increased work during service availabilities. The significance of the SSN aging curve can be observed from the historical trends. As the aging SSNs enter their last maintenance periods ahead of their last operational cycles, the risk for increased work is high. Therefore, the risk for increases in naval shipyard workload and late completion of the last maintenance availabilities for these submarines will increase during the next few years.

    To date, only one nuclear-powered aircraft carrier, the USS *Enterprise* (CVN-65), has reached the end of planned service life. For a variety of reasons, the *Enterprise* was assigned to private-sector main-

Figure 4.2
Percentage of Shipyard-Maintained Vessels Late in Their Service Lives, by
Type, FYs 1980–2040

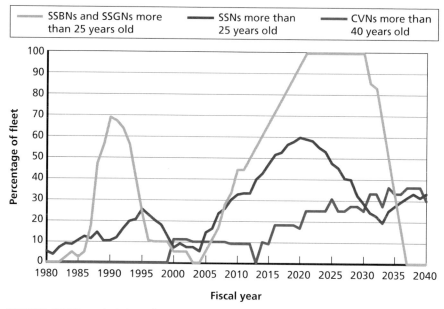

SOURCE: RAND analysis based on U.S. Department of the Navy, *Department of the
Navy Fiscal Year (FY) 2007 Budget Estimates Submission: Justification of
Estimates—Operation and Maintenance, Navy*, Vol. 1, 01-A Exhibit, February 2006.
RAND RR1552-4.2

tenance in a Remaining Service Life Program, executed by Newport
News Shipbuilding. As the *Nimitz*-class carriers age, the number of
carriers with more than 40 years of service life will increase from one
to three and will remain at three for the next four decades. Put another
way, public shipyards, which have never had to provide service for a
carrier with more than 40 years of service, will soon be doing so for
three such carriers—two more such carriers than any facility has ever
had to service.

The planned maintenance resulting from the advanced age
of these ships is significant, and there is evidence that the resulting
unplanned maintenance could also be significant. The effects of aging
are difficult to isolate, particularly from the effects of increased opera-

tions, but maintainers and planners with whom we spoke described specific types of work that were unexpected and were likely associated with both aging and increased operational deployments.

The Navy's maintenance planning activities adjust maintenance planning for age. The planning activities indicate that more maintenance is required for older platforms. For carriers, there is an additional 30–percentage point increase in maintenance man-days planned (and budgeted) in the life-cycle maintenance plan for a newer carrier (as shown in Figure 2.7). Compared with a carrier that has 25 years of service life, a new carrier will receive 15 percent fewer man-days for the same availability, and a carrier with 40 years will receive 15 percent more man-days. Accordingly, there is a direct man-day and budget increase as vessels age. More problematic will be the unplanned maintenance and the increased risk in execution of the prescribed maintenance within the notional availability durations. There is significant risk during such periods for delayed delivery of submarines and carriers due to unforeseen growth, increased planned maintenance, or unpredicted maintenance. This risk affects the critical path for the availabilities and can require increased engineering or testing.

Increased Operational Tempo

In the past, the deployment length for an aircraft carrier was notionally six months, with a 2.0 turnaround ratio (meaning that for every month the carrier was deployed, it needed to spend two months at home). In the 18-month cycle, the 24-month cycle, and the 32-month cycle, each carrier would perform a single six-month deployment. In the 32-month cycle, the carriers would occasionally do a mini-deployment, typically three months long, to cover areas of operation as needed.

To address changes in fleet size and operational demands, deployment length has increased and the turnaround ratio has been reduced, with maintenance-cycle length increasing to 36 months. To accommodate operational demands with a smaller carrier-force structure, carrier deployments are being planned for between eight and ten months, with some even reaching a full year.

The Carrier Planning Activity has collected data on the number of days deployed and the delay to completion of the next DPIA. Its hypothesis is that longer deployments are resulting in significant additional unplanned work, and more time is therefore required to complete the work. Figure 4.3 shows the changes in deployment lengths and DPIA completion schedules for availabilities completed from FYs 2002 through 2014. The average deployment length for aircraft carriers from FYs 2009 through 2014 was 284 days, an increase of 34 percent from the average observed from FYs 2002 through 2008.[1] CVN-69, CVN-72, and CVN-74 all had lengthy deployments and DPIAs that were completed three or more months late. CVN-68,

Figure 4.3
Carrier Deployment Lengths and Completion Dates of Next Availability, FYs 2002–2014

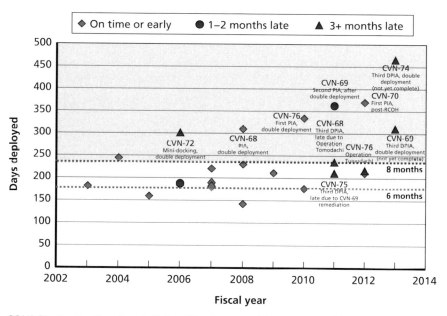

SOURCE: Carrier Planning Activity, "Deployment History," spreadsheet, U.S. Department of the Navy, provided to RAND by Carrier Planning Activity, February 10, 2014a.
NOTE: Year shown is the year of CNO availability following the previous deployment.
RAND RR1552-4.3

---

[1]   Carrier Planning Activity, 2014b.

CVN-70, and CVN-76 had lengthy deployments but maintenance that was completed on time. So what was different between these two groups of availabilities?

The availabilities that were completed on time were on ships that were either newer or had recently completed an RCOH, which essentially resets the ship. Prior to its FY 2010 PIA, CVN-76 had deployed for more than 325 days, but the PIA was the first of the new ship's life, so the carrier was in better shape than other older carriers. Similarly, CVN-70 experienced nearly 375 days deployed prior to its PIA in FY 2012, but that was its first PIA after its RCOH; therefore, like CVN-76, the ship was in better shape than some of the other older carriers. CVN-68 was headed into a modified DPIA, nicknamed a "Super DPIA," following its FY 2008 PIA, so maintenance was deferred into that larger availability.

Several ships did not finish maintenance on time. CVN-69 had a double deployment, a PIA, and then another double deployment. CVN-72 also had a double deployment and a modified PIA, which included some dry-dock work. In FY 2014, CVN-74 and CVN-69 were in the midst of lengthy availabilities following double deployments. The maintenance for these CVNs had not been completed by the time the analysis for this report was complete.

We could not find substantial analysis of the potential near-term and life-cycle effects resulting from the shift to significantly longer carrier deployments. With the higher level of operations, the carrier may need refueling sooner than at the 25-year mark, and the end of life may come sooner than 50 years because of the increased use of nuclear fuel. Maintainers and maintenance planners we interviewed described certain types of work that they believed to be associated with increased operational tempo and aging. These included greater-than-expected increases to shafting repairs and unplanned removal of rudders because of excessive corrosion, among other tasks. Based on discussions with staff from the Carrier Planning Activity and naval shipyards, the additional maintenance generated by the higher operational trends is recoverable but may require longer aircraft carrier dry-docking periods, additional maintenance man-days, and some unplanned component replacements. These observations are a good starting point for quantifying and understanding the effects of increased deployments on maintenance.

Maintenance Planning During Acquisition Development and Production

The maintenance requirements for a new class of ship are established well in advance of performing the first maintenance activity. The requirements are often based on the best information available at that time. As a vessel moves from development to production to operation (and thereby enters the fleet), better information about maintenance requirements becomes available. The incentives to keep life-cycle costs within affordability caps, however, can lead to overly optimistic estimates for maintenance during the acquisition process, and then maintenance budgets change to support more-realistic requirements when the platform enters the fleet.

Figure 4.4 illustrates the evolution of the Class Maintenance Plan maintenance man-days for a single *Virginia*-class submarine. The man-days presented are the man-days expected to be required over the life of the ship. The dramatic increase in prescribed maintenance reflects the submarine's transition from acquisition to operation. An aggressive notional maintenance assumed in the acquisition phase of the *Virginia* class was adjusted when the lead submarines entered the fleet, resulting in changes to both naval shipyard budgeting and loading. At the time of this research, there have been only three major maintenance periods and one extended dry-docking selected restricted availability for *Virginia*-class submarines. Additional learning will occur as new activities are performed, which presents additional risk to the future budget.

Figure 4.4
Life-Cycle Maintenance Requirements for a *Virginia*-Class Submarine,
Selected Years

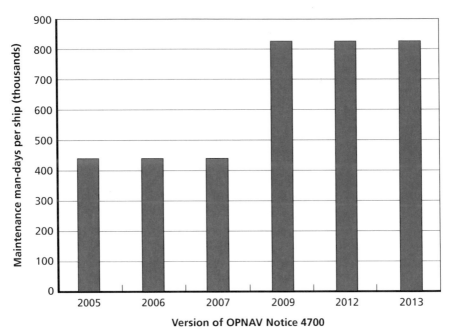

SOURCE: OPNAV Notice 4700, various years.
NOTE: The RAND study team did not have data for 2008, 2010, and 2011.
RAND RR1552-4.4

Estimating Risk

Recognizing that risks need to be addressed prior to execution, NAVSEA assesses risk to the performance of availabilities based on a defined list of considerations.[2] These include the following:

- past performance on the same type of availability (future performance is expected to be similar to past performance)
- the number of key events occurring simultaneously

---

[2] NAVSEA 04, "Chief of Naval Operations Submarine Performance Factor Assessments," Standard Operating Procedure No. 13 Revised, undated, provided to RAND by NAVSEA 04 in September 2014.

- the complexity of the availability, which is based on the number of man-days per month required to execute the availability within the allocated time
- the availability of resources (if there are inadequate personnel with the needed trade-skill, then the resources are considered constrained)
- priority of the hull (OPNAV policy[3] dictates the priority of execution of shipyard work).

There are specific rules to determine a category of risk from 1 (lowest) to 5 (highest). If certain types of availabilities overlap, then NAVSEA indicates a higher risk category, especially as the length of overlap increases. The mathematical estimation of the risk is referred to as the performance factor and is a number between 0 and 1. Typically, the value is above 0.85. If an availability is estimated to require 100,000 total man-days, with a performance factor of 0.98, then the shipyard will be given approximately 102,000 man-days to complete it.[4] The 102,000 man-days are then priced and put in the Program Objective Memorandum (POM) as the requirement for the availability. Similarly, an availability planned for 100,000 total man-days but having a performance factor of 0.94 would be given roughly 106,000 man-days to execute. During the process of establishing the performance factor, negotiations can end up increasing the factor to a level higher than what the technical analysis yields.

Table 4.2 illustrates the planned, approved (or budgeted), and actual performance factors for public shipyard maintenance. The planned performance factor is the average of the performance factors programmed in the POM for the availability indicated. The approved performance factor is the average of what the fleet directs the shipyards to execute in guidance, and it is almost always the same as the planned factor. The actual performance factor is the average of what was actually executed.

---

[3] See OPNAV Instruction 4700.7L, *Maintenance Policy for United States Navy Ships*, Office of the Chief of Naval Operations, U.S. Department of the Navy, May 25, 2010.

[4] $100,000/0.98 = 102,041$.

Table 4.2
Average Planned, Approved, and Actual Performance Factors for Maintenance Executed at Public Shipyards

Ship	Availability	Average Performance Factor		
		Planned	Approved	Actual
Nimitz class (CVN-68)	DPIA	0.94	0.94	0.86
Nimitz class (CVN-68)	PIA	0.97	0.97	0.93
Nimitz class (CVN-68)	Selected restricted availability (forward-deployed naval force)	0.99	0.99	0.99
Los Angeles class (SSN-688)	Dry-docking selected restricted availability	0.94	0.94	0.92
Los Angeles class (SSN-688)	Engineered overhaul	0.92	0.92	0.92
Virginia class (SSN-774)	Extended dry-docking selected restricted availability	0.88	0.88	0.85
Ohio class (SSBN-726)	ERO	0.96	0.94	0.89

Submarines in the *Virginia* class (SSN-774) have a planned performance factor of 0.88, showing that there is significant risk in execution assumed for this class. Indeed, actual performance to date is somewhat lower, at 0.85, showing an even higher risk than was assumed. Because the numbers of *Virginia*-class submarines will increase in coming years, as will extended dry-docking selected restricted availabilities for them, we expect identification and execution of and lessons learned from such work to increase naval shipyard work.

Performance factors for both *Nimitz*-class (CVN-68) DPIAs and *Ohio*-class (SSBN-726) EROs reflect significant challenges, which we discussed in Chapter Two. For example, the planned performance factor for the CVN DPIA is 0.94, but the average actual factor is 0.86. Given previously mentioned challenges on ship aging, these availabilities may continue to cause increases in naval shipyard workload and resource requirements.

Summary

Estimating future workload requirements involves considering many risks that may not be easily characterized. Many maintenance activities are emergent, and those that are planned can ultimately vary significantly from the planned values. The requirements are based on assumptions about how the platform will operate, and these assumptions may change over the life of the vessel. The acquisition, budgeting, and availability-planning processes can increase the uncertainty. For example, incentives for keeping life-cycle costs at a minimum during the acquisition process can lead to overly optimistic maintenance plans. The budgeting process requires the Navy to estimate maintenance requirements two years prior to execution, when much of the maintenance activity is still unknown. And the POM process underfunds maintenance requests to varying degrees across submarines, carriers, and surface ships. Furthermore, the efforts to estimate risk and plan for the unexpected may be thwarted by overly optimistic performance factors. All of this points to a risk of additional, unplanned work for the public shipyards in the execution year. The Navy should consider studies to address these risks and to better inform the development of the next POM.

CHAPTER FIVE

Summary, Conclusions, and Recommendations

Since FY 2004, the direct and indirect man-days executed at the Navy's public shipyards have increased and are expected to increase further. From FYs 2004 through 2014, the direct man-days executed rose by 7.2 percent, and the civilian workforce rose by 17.4 percent (see Figure 5.1).

For a variety of reasons, the direct man-day increases have occurred primarily in SSBN and CVN CNO availabilities and in scheduled maintenance. The maintenance requirements for SSBNs have increased by 13 percent. These additional requirements are a result of additional new work, unplanned repairs, execution performance, resource constraints, and work-prioritization policies. On occasion, resources are diverted from the SSBN ERO to complete other availabilities, which increases the schedule of the ERO and ultimately adds cost. Some examples of this practice are shown in Table 5.1.

The increases in CVN work have occurred because there are more nuclear-powered carriers being maintained by the naval shipyards. Even though the overall carrier fleet has decreased, the number of CVNs being maintained at the shipyards will have increased by 38 percent between FYs 2004 and 2016 (see Figure 5.2). In addition, increases to deployment cycles and changes to the life-cycle maintenance plans have led to more workload needing to be accomplished in fewer maintenance periods, creating fluctuations in demand that are difficult for the shipyards to manage (see Figure 5.3).

Figure 5.1
Civilian Staffing Levels and Direct Man-Days Executed at Public Shipyards,
FYs 2004–2014

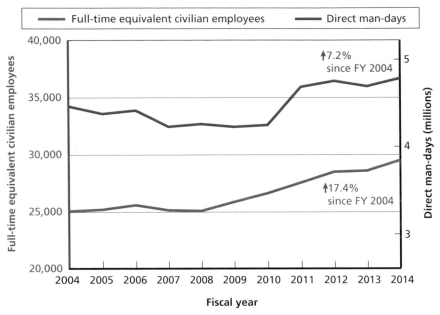

SOURCE: RAND analysis based on Norfolk Naval Shipyard, 2014; Pearl Harbor Naval Shipyard, 2014; Portsmouth Naval Shipyard, 2014; and Puget Sound Naval Shipyard, 2014.
RAND RR1552-5.1

Scheduled maintenance has increased as more of the life-cycle maintenance work is pushed to continuous maintenance and fleet maintenance periods. Indirect man-days have increased as a result of increases in direct man-days and in training. Full understanding of the causes of increases to indirect man-days would require more data than was available for this analysis.

The number of direct man-days executed in FY 2010 was around 4.3 million. The estimated direct man-days planned for execution in FY 2018 are 5.5 million—a 28-percent increase from FY 2010 levels. The planned increase in civilian personnel from FYs 2010 through 2018 is nearly 7,000 people—a 26-percent increase. Executing the plan

Table 5.1
Changes to Engineered Refueling Overhaul Availabilities for Nuclear-Powered Ballistic-Missile Submarines

Ship	Rationale	Shipyard Effect
USS *Alaska* (SSBN-732)	Extension of CNO end date	+ 16 days
USS *Tennessee* (SSBN-734)	Shipyard performance	+ nearly 2 months
	Change in end date	+ 2 months
USS *Pennsylvania* (SSBN-735)	Extension of CNO end date	+ 1 month
	Unplanned nuclear valve repairs and elimination of required 30-day schedule buffer	+ 3 months
USS *West Virginia* (SSBN-736)	Resource-constrained reschedule	+ 55 days
	Additional new work, performance reduction factor	+ 18,050 man-days

to raise naval shipyard manning levels to 35,500 civilians is necessary but not sufficient to meet this near-term need. The increasingly inexperienced workforce has created a near-term productivity deficit, with the effective workforce being less than planned.

Even with accelerated hiring and accelerated training, our productivity model predicts that the effective workforce will not be sufficient for the peak naval shipyard workload planned in FYs 2016–2017 (as shown in Figure 3.5). The suggested and currently planned resourcing plan may result in excess labor in the middle term, with a recapitalized and trained workforce and a decrease in predicted workload around FY 2022.

Additional work resulting from known and unknown risks, however, may make near-term peaks unexecutable and may drive future workload above predicted levels. If historical performance is any indication of the future, the peak workload could exceed 6.6 million total man-days, which would be a 55-percent increase over the FY 2010 levels (as shown in Figure 4.1).

Figure 5.2
Changes to Nuclear-Powered Aircraft Carrier Fleets Maintained at Public Shipyards

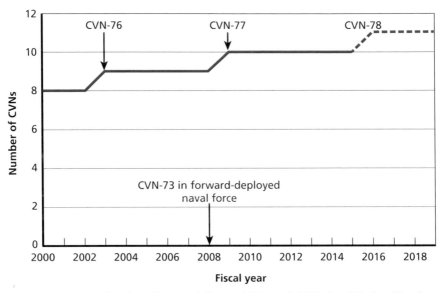

SOURCE: RAND analysis based on Norfolk Naval Shipyard, 2014; Pearl Harbor Naval Shipyard, 2014; Portsmouth Naval Shipyard, 2014; and Puget Sound Naval Shipyard, 2014.
NOTE: This figure does not include non-nuclear carriers or CVN-65, which was maintained at Newport News during this period.
RAND RR1552-5.2

In addition to the current resourcing plan, the Navy should consider the following options:

- Hire additional resources. The productivity model, historical trends, and risk factors predict a need above currently planned levels.
- Continue investment in accelerated training programs across all critical trades and find new hires that have experience. Accelerated training creates productive workers more quickly and minimizes the potential increases to the size of the workforce. Hiring individuals with even one year of experience can have a significant effect on overall workforce productivity.

Figure 5.3
Changes to Nuclear-Powered Aircraft Carrier Workloads at Public Shipyards

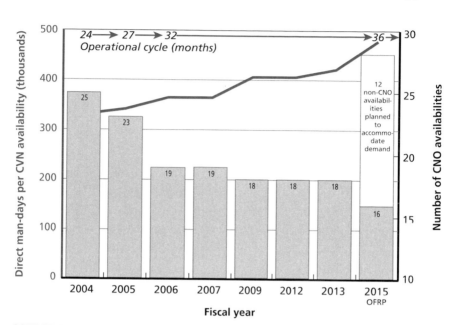

SOURCE: RAND analysis based on Norfolk Naval Shipyard, 2014; Pearl Harbor Naval Shipyard, 2014; Portsmouth Naval Shipyard, 2014; and Puget Sound Naval Shipyard, 2014.
NOTE: This figure does not include non-nuclear carriers or CVN-65, which was maintained at Newport News during this period. No data were found for FYs 2008, 2010, 2011, and 2014.
RAND *RR1552-5.3*

- Mitigate workload. A fully productive workforce cannot be established in time to execute the near-term peaks. As a result, the Navy should consider outsourcing some of the work.

The Navy needs to develop a more robust workforce-forecasting capability to assess and predict naval shipyard personnel requirements. There is too much uncertainty affecting personnel requirements. The Navy should have a forecasting capability that can incorporate workforce productivity and risk. Many of the variables used to estimate future resource requirements, such as retention rates, attrition rates, and expected future workload, vary. This variability should be considered when estimating future personnel needs.

The number of nuclear platforms that will be in the latter parts of their expected service lives in the next few years is unprecedented (as shown in Figure 4.2). The implications of this are not well known. The Navy should invest in studies to better understand and program for the effects of this aging.

Finally, to better manage fluctuations in demand for maintenance, and to more efficiently use its limited resources, the Navy should conduct a strategic, long-term evaluation of the role of the private sector, including shipbuilders, in life-cycle (including IRR) maintenance and modernization of all platforms managed by the naval shipyards. The Navy should look at the future maintenance work as a portfolio of work, with alternative paths for managing that work, each of which should be evaluated. The allocation of public shipyard work to the private sector is currently done mostly on an ad-hoc basis. When there is a peak that the public shipyards cannot execute, as there will be in FY 2017, the Navy turns to the private sector for help. A better long-term strategic plan could allow for better planning and potential cost savings.

Sensitivity Analysis on Productivity Assumption

The amount of time it takes to become fully productive varies by trade. While we assume that, on average, it takes seven years to become fully productive, there are some trades in which apprentices may become journeymen more quickly, and others less quickly. Figure A.1 shows how the assumption for the time it takes to become fully productive affects the number of civilian personnel needed. The longer it takes to become fully productive, the more civilian personnel who must be hired to accomplish the work. If every new hire took one year to become fully productive, the predicted hiring need would be thousands less than if full productivity took four or seven years.

Figure A.1
Predicted Number of Civilian Personnel Needed, by Timetable to Be Fully Productive

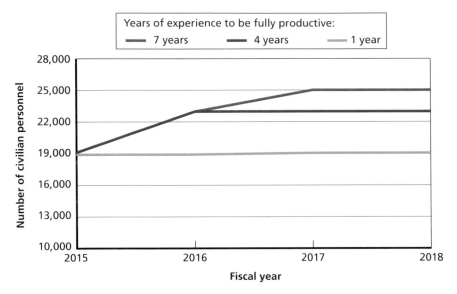

SOURCE: RAND analysis based on Norfolk Naval Shipyard, 2014; Pearl Harbor Naval Shipyard, 2014; Portsmouth Naval Shipyard, 2014; and Puget Sound Naval Shipyard, 2014.
RAND RR1552-A.1

Bibliography

Assistant Deputy Commander, Industrial Operations, *Chief of Naval Operations Submarine Performance Factor Assessments*, Standard Operating Procedure No. 13, Washington, D.C.: U.S. Department of the Navy, undated.

Bechet, Thomas, *Strategic Staffing*, New York: American Management Association International, 2002.

Carrier Planning Activity, *Submarine Engineered Operating Cycle Program*, OPNAVINST 3120.33C N9, Washington, D.C.: U.S. Department of the Navy, January 22, 2013.

———, "Deployment History," spreadsheet, U.S. Department of the Navy, provided to RAND by Carrier Planning Activity, February 10, 2014a.

———, "Master Avail Date List," spreadsheet, U.S. Department of the Navy, provided to RAND by Carrier Planning Activity, March 27, 2014b.

———, "Historic CV and CVN Deployments," spreadsheet, U.S. Department of the Navy, provided to RAND by Carrier Planning Activity, June 19, 2014c.

Commander, Norfolk Naval Shipyard, "Re-Baseline Request for USS WASP (LHD-1) FY08 Docking Phased Maintenance Availability," Norfolk, Va.: U.S. Department of the Navy, February 17, 2009a.

———, "Re-Baselining Request for USS *Alaska* (SSBN-732) FY07 Engineered Refueling Overhaul," Norfolk, Va.: U.S. Department of the Navy, May 15, 2009b.

———, "Re-Baselining Request for USS *Boise* (SSN-764) FY09 Drydocking Selected Restricted Availability (DSRA), and USS *Frank Cable* (AS-40) FY09 PHASED Maintenance Availability (PMA)," Norfolk, Va.: U.S. Department of the Navy, May 21, 2009c.

———, "Second Re-Baselining Request for USS *Kearsarge* (LHD-3) FY09 Docking Phased Maintenance Availability," Norfolk, Va.: U.S. Department of the Navy, May 22, 2009d.

———, "Re-Baselining Request for USS *Abraham Lincoln* (CVN-72) FY09 Planned Incremental Availability," Norfolk, Va.: U.S. Department of the Navy, July 24, 2009e.

———, "Re-Baseline Request for Multiple Availabilities at Norfolk Naval Shipyard," Norfolk, Va.: U.S. Department of the Navy, February 14, 2013a.

———, "Supplemental Letter to NNSY's Multiple Availability Rebaselining Request for the USS *West Virginia* Engineered Refueling Overhaul," Norfolk, Va., U.S. Department of the Navy, June 6, 2013b.

———, "Rebaseline Request for USS *Newport News* (SSN 750) Fiscal Year 2012 Engineered Overhaul (EOH)," Norfolk, Va., U.S. Department of the Navy, September 14, 2013c.

———, "Rebaselining Request for USS *Albany* (SSN 753) Fiscal Year 2014 Engineered Overhaul," Norfolk, Va., U.S. Department of the Navy, March 7, 2014a.

———, "Rebaselining Request for USS *Maryland* (SSBN 738) Fiscal Year 2013 Engineered Refueling Overhaul," Norfolk, Va., U.S. Department of the Navy, March 7, 2014b.

———, "Rebaselining Request for Multiple Availabilities at Norfolk Naval Shipyard," Norfolk, Va., U.S. Department of the Navy, June 20, 2014c.

Commander, Pearl Harbor Naval Shipyard and Intermediate Maintenance Facility, "Re-Baselining Request for USS *Minneapolis Saint Paul* (SSN 708) FY07 IA, USS *Key West* (SSN 722) FY08 DSRA, and USS *Houston* (SSN 713) FY08 DSRA," Pearl Harbor, Hawaii, U.S. Department of the Navy, March 27, 2008.

———, "Re-Baseline Change Request to Pearl Harbor Naval Shipyard and Intermediate Facility (PHNS&IMP) FY13 Execution Guidance for USS *Columbia* (SSN771) Docking Selected Availability (DSRA)," Pearl Harbor, Hawaii, U.S. Department of the Navy, November 13, 2012.

———, "Re-Baseline Change Request to Pearl Harbor Naval Shipyard and Intermediate Facility (PHNS&IMP) FY13 Execution Guidance for USS *Columbia* (SSN 771), USS *Louisville* (SSN 724) and USS *Greeneville* (SSN 772) Docking Selected Restricted Availability (DSRA)," Pearl Harbor, Hawaii, U.S. Department of the Navy, February 8, 2013.

———, "Re-Baseline Change Request to Pearl Harbor Naval Shipyard and Intermediate Maintenance Facility (PHNS&IMF) FY14 Execution Guidance for Overtime," Pearl Harbor, Hawaii, U.S. Department of the Navy, April 11, 2014a.

———, "Re-Baseline Change Request to Pearl Harbor Naval Shipyard and Intermediate Maintenance Facility (PHNS&IMF) FY14 Execution Guidance for USS *Texas* (SSN 775) Extended Docking Selected Restricted Availability (EDSRA), USS *Cheyenne* (SSN 773) Docking Selected Restricted Availability (DSRA) and USS *Tucson* (SSN 770) DSRA," Pearl Harbor, Hawaii, U.S. Department of the Navy, May 8, 2014b.

————, "Re-Baseline Change Request to Pearl Harbor Naval Shipyard and Intermediate Maintenance Facility (PHNS&IMF) FY14 Execution Guidance for USS *Bremerton* (SSN 698) Docking Selected Restricted Availability (DSRA), USS *Cheyenne* (SSN 773) DSRA, USS *Tuscon* (SSN 770) DSRA and Overtime," Pearl Harbor, Hawaii, U.S. Department of the Navy, May 30, 2014c.

————, "Re-Baseline Change Request to Pearl Harbor Naval Shipyard and Intermediate Maintenance Facility (PHNS&IMF) FY14 Execution Guidance for USS *Texas* (SSN 775) Extended Docking Selected Restricted Availability (EDSRA), and USS *Buffalo* (SSN 715) Pre-Inactivation Restricted Availability (PIRA)," Pearl Harbor, Hawaii, U.S. Department of the Navy, August 11, 2014d.

————, "Re-Baseline Change Request to Pearl Harbor Naval Shipyard and Intermediate Maintenance Facility (PHNS&IMF) FY14 Execution Guidance for USS *Bremerton* (SSN 698) Docking Selected Restricted Availability (DSRA), and USS *Cheyenne* (SSN 773) DSRA," Pearl Harbor, Hawaii, U.S. Department of the Navy, August 25, 2014e.

Commander, Portsmouth Naval Shipyard, "Re-Baseline Change Request to Portsmouth Naval Shipyard FY08 Execution Guidance," Portsmouth, N.H., U.S. Department of the Navy, April 30, 2008.

————, "Re-Baseline Change Request to Portsmouth Naval Shipyard Fiscal Year 2014 Execution Guidance (USS *Alexandria* (SNN 757) Engineered Overhaul)," Portsmouth, N.H., U.S. Department of the Navy, February 3, 2014a.

————, "Re-Baseline Change Request to Portsmouth Naval Shipyard Fiscal Year 2014 Execution Guidance (USS *Dallas* (SNN 700) Pre-Inactivation Restricted Availability)," Portsmouth, N.H., U.S. Department of the Navy, February 10, 2014b.

————, "Re-Baseline Change Request to Portsmouth Naval Shipyard Fiscal Year 2014 Execution Guidance (USS *Albuquerque* (SNN 706) Pre-Inactivation Restricted Availability)," Portsmouth, N.H., U.S. Department of the Navy, April 1, 2014c.

Commander, Puget Sound Naval Shipyard and Intermediate Maintenance Facility, "Re-Baseline for Puget Sound Naval Shipyard and Intermediate Maintenance Facility Fiscal Year 2014 Execution Guidance," Bremerton, Wash., U.S. Department of the Navy, October 16, 2014.

Commander, Submarine Force (N43), "USS *Boise* (SSN 764) FY09 Docking Selected Restricted Availability (DSRA) TYCOM Final Review Estimate (FRE) Acceptance," Norfolk, Va., U.S. Department of the Navy, March 23, 2009.

Fleet Maintenance Board of Directors, *Program Objective Memorandum-15 (POM15) Ship Maintenance 9-Step Process Guide and Schedule*, Norfolk, Va., U.S. Department of the Navy, August 2, 2012.

———, *Program Objective Memorandum-17 (POM15) Ship Maintenance 9-Step Process Guide and Schedule*, Norfolk, Va., U.S. Department of the Navy, August 2, 2012.

———, *Program Objective Memorandum-17 (POM17) Ship Maintenance 9-Step Process Guide and Schedule*, Norfolk, Va., U.S. Department of the Navy, August 8, 2014.

Institute for Work and Health, "What Researchers Really Mean by . . . Statistical Significance," *At Work*, No. 40, Spring 2005. As of October 11, 2016: http://www.iwh.on.ca/wrmb/statistical-significance

McDermott, Deborah, "Sequestration Threatens Portsmouth Naval Shipyard Jobs, Workload," *Seacoast Online*, September 20, 2012. As of October 24, 2016: http://www.seacoastonline.com/article/20120920/NEWS/209200420

Naval Sea Systems Command, *Industrial Ship Safety Manual for Submarines*, s9002-AK-CCM-010/6010, Washington, D.C.: U.S. Department of the Navy, January 8, 2009a.

———, *Uniform Costing Policies and Cost Classes Procedures for Mission Funded Naval Shipyards*, Washington, D.C.: U.S. Department of Navy, March 13, 2009b.

———, "Approval of and Initial Promulgation of the Industrial Ship Safety Manual for Fire Prevention and Response," February 6, 2014.

———, *Strategic Business Plan, 2013–2018*, 2nd ed., Washington, D.C., 2015. As of September 27, 2016: http://www.navsea.navy.mil/Portals/103/Documents/Strategic%20Documents/ SBP13-14_Final-2ndEd.pdf

Naval Sea Systems Command, Logistics, Maintenance, and Industrial Operations Directorate, "Chief of Naval Operations Submarine Performance Factor Assessments," Standard Operating Procedure No. 13 Revised, undated, provided to RAND by NAVSEA 04, September 2014.

NAVSEA 04—*See* Naval Sea Systems Command, Logistics, Maintenance, and Industrial Operations Directorate.

Norfolk Naval Shipyard, "WF-300 Workload Allocation and Resource Report (WARR)," spreadsheet, provided to RAND by NAVSEA 04, July 2014.

OPNAV Instruction 3120.33C, *Submarine Engineered Operating Cycle Program*, Office of the Chief of Naval Operations, January 22, 2013.

OPNAV Instruction 4700.7L, *Maintenance Policy for United States Navy Ships*, Office of the Chief of Naval Operations, U.S. Department of the Navy, May 25, 2010. As of October 24, 2016: http://www.navybmr.com/study%20material/OPNAVINST%204700.7L.pdf

OPNAV Notice 4700, *Representative Intervals, Durations, Maintenance Cycles, and Repair Man-Days for Depot Level Maintenance Availabilities of U.S. Navy Ships*, Office of the Chief of Naval Operations, U.S. Department of the Navy, various years.

Pearl Harbor Naval Shipyard, "WF-300 Workload Allocation and Resource Report (WARR)," spreadsheet, provided to RAND by NAVSEA 04, July 2014.

Portsmouth Naval Shipyard, "WF-300 Workload Allocation and Resource Report (WARR)," spreadsheet, provided to RAND by NAVSEA 04, July 2014.

Public Law 112-25, Budget Control Act of 2011, August 2, 2011.

Puget Sound Naval Shipyard, "WF-300 Workload Allocation and Resource Report (WARR)," spreadsheet, provided to RAND by NAVSEA 04, July 2014.

Riposo, Jessie, Brien Alkire, John F. Schank, Mark V. Arena, James G. Kallimani, Irv Blickstein, Kimberly Curry Hall, and Clifford A. Grammich, *U.S. Navy Shipyards: An Evaluation of Workload- and Workforce-Management Practices*, Santa Monica, Calif.: RAND Corporation, MG-751-NAVY, 2008. As of October 24, 2016:
http://www.rand.org/pubs/monographs/MG751.html

Schank, John F., Mark V. Arena, Paul DeLuca, Jessie Riposo, Kimberly Curry, Todd Weeks, and James Chiesa, *Sustaining U.S. Nuclear Submarine Design Capabilities*, Santa Monica, Calif.: RAND Corporation, MG-608-NAVY, 2007. As of September 27, 2016:
http://www.rand.org/pubs/monographs/MG608.html

Welch, Larry D., and John C. Harvey, Jr., *Independent Review of the Department of Defense Nuclear Enterprise*, Washington, D.C.: U.S. Department of Defense, June 2, 2014.

Yardley, Roland J., John F. Schank, James G. Kallimani, Raj Raman, and Clifford A. Grammich, *A Methodology for Estimating the Effect of Aircraft Carrier Operational Cycles on the Maintenance Industrial Base*, Santa Monica, Calif.: RAND Corporation, TR-480-NAVY, 2007. As of October 20, 2016:
http://www.rand.org/pubs/technical_reports/TR480.html

U.S. Department of the Navy, "Budget Materials," web page, Washington, D.C., various years. As of October 7, 2016:
http://www.secnav.navy.mil/fmc/fmb/Pages/Fiscal-Year-2017.aspx

———, *Department of the Navy Fiscal Year (FY) 2007 Budget Estimates Submission: Justification of Estimates—Operation and Maintenance, Navy*, Vol. 1, 01-A Exhibit, February 2006.